Evaluating Climate Change Governance in Bangladesh

Md. Faruk Hossain · Salim Momtaz

Evaluating Climate Change Governance in Bangladesh

Springer

Md. Faruk Hossain
Department of Geography
and Environment
University of Dhaka
Dhaka, Bangladesh

Salim Momtaz
School of Environmental
and Life Sciences
University of Newcastle
Sydney, NSW, Australia

ISBN 978-3-031-98485-3 ISBN 978-3-031-98486-0 (eBook)
https://doi.org/10.1007/978-3-031-98486-0

This Springer imprint is published by the registered company Springer Nature Switzerland AG
The registered company address is: Gewerbestrasse 11, 6330 Cham, Switzerland

Preface

Climate change poses a significant concern for Bangladesh and its population. Climate change governance practices span from national to local levels, aiming to mitigate adverse effects. These practices enhance resilience, improving the lives and livelihoods of vulnerable communities. Our book evaluates climate change governance by governmental and non-governmental organizations in Bangladesh.

Through an extensive literature review, we gained insights into the research issue. Assessing climate change governance involved five key criteria: accountability, transparency, participation, government effectiveness, and resource allocation. Employing a mixed-method research design, we interpreted findings using an interpretative phenomenological approach. Data collection included surveys, interviews, and focus group discussions in study areas, supplemented by secondary sources.

Our study reveals collaborative efforts between governmental and non-governmental organizations over the past two decades. Despite limitations, they ensure transparency and stakeholder engagement in climate-related projects. Climate change governance remains crucial, and individuals adopt diverse strategies supported by these organizations to overcome challenges. While political commitment exists, Bangladesh faces ongoing governance challenges. Drawing from our research, we propose recommendations to enhance climate change governance practices in the country.

Dhaka, Bangladesh — Md. Faruk Hossain
Sydney, Australia — Salim Momtaz

Acknowledgements We extend our heartfelt gratitude to the University of Newcastle, Australia, for funding this study. We sincerely appreciate the cooperation of various officers from governmental and non-governmental organizations in Bangladesh during our fieldwork. Additionally, we acknowledge the local individuals who generously shared their experiences through household surveys, key informant interviews, and focus group discussions. Our thanks also go to the local authorities who facilitated and supported our work in the field.

Contents

1 Introduction and Background 1
- 1.1 Introduction 1
- 1.2 Study Context and Significance 3
- 1.3 Study Objectives 4
- 1.4 Methodology and Approaches 5
- 1.5 Book Structure and Content 6
- References 7

2 Literature Review—Climate Change Governance in Bangladesh—Perspectives of Governmental and Non-governmental Organisations 11
- 2.1 Climate Change Governance—Theoretical Considerations 11
- 2.2 Effective Climate Change Governance 11
- 2.3 Climate Change Governance Stakeholders 13
- 2.4 Climate Change and Sustainable Development 16
- 2.5 Climate Change Governance Status in Developing Countries 17
- 2.6 Climate Change Governance Status in Bangladesh 21
- 2.7 Climate Change Governance Evaluation Criteria 23
- 2.8 Climate Change Finance 25
- 2.9 Climate Change Governance Evolution in Bangladesh 27
- References 30

3 Methodology 37
- 3.1 Introduction 37
- 3.2 Research Methods 37
- 3.3 Research Design 38
  - 3.3.1 Triangulation 39
- 3.4 The Study Area 39
  - 3.4.1 Structure of Administrative Units in Bangladesh 40
  - 3.4.2 Study Area Details 40
  - 3.4.3 Hatiya Upazila at a Glance 42
  - 3.4.4 Disasters and Climate Events in Hatiya Upazila 45

3.4.5 Researcher's Observation of the Study Area 49
3.5 Data Collection Techniques 50
3.5.1 Document Review (Secondary Data and Information Collection) 51
3.5.2 Primary Data and Information Collection 51
3.5.3 Key Informant Interviews 53
3.5.4 Focus Group Discussions 54
3.6 Data Analysis 55
3.6.1 Quantitative Data 56
3.6.2 Qualitative Data 57
3.7 Reliability and Validity 59
3.7.1 Validity 59
3.7.2 Reliability 59
3.8 Ethical Considerations 60
3.9 Procedures Used 61
3.10 Conclusion 61
References 62

4 Climate Change Governance Practices of Governmental Organisations in Bangladesh 65
4.1 Introduction 65
4.2 Governmental Organisations in Climate Change-Related Activities 65
4.2.1 Ministry of Environment, Forest and Climate Change 66
4.2.2 Department of Environment 66
4.2.3 Climate Change Cell 68
4.2.4 Bangladesh Climate Change Trust 68
4.2.5 Implementation, Monitoring and Evaluation Division 72
4.2.6 Department of Disaster Management 73
4.2.7 Non-governmental Organisation Affairs Bureau 73
4.2.8 Palli Karma-Sahayak Foundation 74
4.3 Plans and Programs on Climate Change Issues 74
4.3.1 Bangladesh Climate Change Strategy and Action Plan 74
4.3.2 National Adaptation Plan of Action 76
4.3.3 Eighth Five Year Plan 2020–2025 77
4.3.4 National Determined Contributions 78
4.3.5 Bangladesh Delta Plan 2100 78
4.4 Institutional Frameworks 79
4.4.1 Strengths of the Institutional Framework 81
4.4.2 Weaknesses of the Institutional Framework 82

4.5 Achievements of Governmental Organisations in Climate Change-Related Activities 83
4.6 Challenges of Governmental Organisations in Climate Change-Related Activities 85
4.7 Governmental Organisation Accountability in Climate Change Activities 89
4.8 Governmental Organisation Transparency in Climate Change Activities 92
4.9 Participation of Governmental Organisations in Climate Change-Related Activities 93
4.10 Effectiveness of Governmental Organisations in Climate Change-Related Activities 95
4.11 Funding Governmental Organisations for Climate Change Activities 96
4.12 Changes in Climate Change Governance 99
4.13 Discussion and Conclusion 101
References 102

5 Climate Change Governance Practices of Non-governmental Organisations in Bangladesh 107
5.1 Introduction 107
5.2 Non-governmental Organisations in Climate Change-Related Activities 107
5.3 Non-governmental Organisation Institutional Arrangements 113
5.4 Non-governmental Organisation Climate Change Governance Practices 115
5.4.1 Non-governmental Organisation Accountability in Climate Change Activities 116
5.4.2 Non-governmental Organisation Transparency in Climate Change Activities 119
5.4.3 Stakeholder Participation in Non-governmental Organisation Activities 123
5.4.4 Non-governmental Organisation Effectiveness in Climate Change Activities 126
5.4.5 Non-governmental Organisation Resource Dependency and Fund Mobilisation Strategies 128
5.5 Non-governmental Organisation Successes in Climate Change-Related Activities 132
5.6 Non-governmental Organisation Challenges in Climate Change-Related Activities 135
5.7 Discussion and Conclusion 139
References 141

6 Community Perceptions of Climate Change Governance 145
6.1 Introduction 145
6.2 Community Age Structure 145
6.3 Educational Level 146
6.4 Marital Status 147
6.5 Household Heads Occupational Status 148
6.6 Respondent House Types 149
6.7 Respondents' Knowledge of Climate Change 150
6.8 Perceptions of Climate Change 151
6.9 Perception of Climate Events 151
6.10 Perceptions of Climate Change Effects 153
6.11 Relationship of Stakeholders in Climate Change Governance 154
6.12 Climate Change Governance Practices at the Local Level 157
6.12.1 Accountability 157
6.12.2 Transparency 159
6.12.3 Government Effectiveness 159
6.12.4 Resources 160
6.12.5 Participation 162
6.13 Overall Climate Change Governance Status at the Local Level 164
6.14 Perceptions About the Effectiveness of Climate Change Governance 165
6.15 Challenges for Local Climate Change Governance 167
6.16 Discussion and Conclusion 171
References 172

7 Conclusion and Recommendations 175
7.1 Summary 175
7.2 Recommendations 177
7.3 Study Contributions 179
7.4 Study Limitations 180
7.5 Future Research Directions 181

Appendix A: Head-of-Household Survey Semi-structured Questionnaire 183

Appendix B: Questionnaire for the Key Informants 191

Appendix C: Questions for Focus Group Discussions 193

Appendix D: Frequently Used Words in KIIs and FGDs (Word Clouds) 195

Abbreviations

ADB	Asian Development Bank
ADP	Annual Development Programme
BARC	Bangladesh Agricultural Research Council
BARI	Bangladesh Agricultural Research Institute
BAU	Bangladesh Agricultural University
BB	Bangladesh Bank
BBS	Bangladesh Bureau of Statistics
BCAS	Bangladesh Centre for Advanced Studies
BCCRF	Bangladesh Climate Change Resilient Fund
BCCSAP	Bangladesh Climate Change Strategy and Action Plan
BCCT	Bangladesh Climate Change Trust
BCCTF	Bangladesh Climate Change Trust Fund
BDP	Bangladesh Delta Plan
BDT	Bangladeshi Taka
BIDS	Bangladesh Institute of Development Studies
BMD	Bangladesh Meteorological Department
BRAC	Bangladesh Rural Advancement Committee
BRRI	Bangladesh Rice Research Institute
BUET	Bangladesh University of Engineering and Technology
BWDB	Bangladesh Water Development Board
CAG	Comptroller and Auditor General
CCA	Climate Change Adaptation
CCC	Climate Change Cell
CCDB	Christian Commission for Development in Bangladesh
CCU	Climate Change Unit
CDMP	Comprehensive Disaster Management Program
CEGIS	Centre for Environmental and Geographic Information Services
CFTM	Climate Finance Transparency Mechanism
CNRS	Centre for Natural Resources Studies
COP	Conference of the Parties
CPP	Cyclone Preparedness Program

CPRD	Centre for Participatory Research and Development
CSO	Civil Society Organisation
DANIDA	Danish International Development Agency
DC	District Commissioner
DDM	Department of Disaster Management
DF	Disaster Forum
DFID	Department for International Development
DM	Disaster Management
DMB	Disaster Management Bureau
DoA	Department of Agriculture
DoE	Department of Environment
DRR	Disaster Risk Reduction
DSSC	Dhaka South City Corporations
DU	Dhaka University
DUS	Dwip Unnayan Songstha (Island Development Association)
ECNEC	Executive Committee of the National Economic Council
ERD	Economic Relations Division
FAO	Food and Agriculture Organization
FD	Forest Department
FGD	Focus Group Discussion
GCF	Green Climate Fund
GDP	Gross Domestic Product
GEF	Global Environment Facility
GOs	Governmental Organisations
IF	Institutional Framework
IIED	International Institute for Environment and Development
IMED	Implementation, Monitoring and Evaluation Division
INGO	International Non-governmental Organisation
IPCC	Intergovernmental Panel on Climate Change
IWFM	Institute of Water and Flood Management
JICA	Japan International Cooperation Agency
KII	Key Informant Interview
LCDP	Low Carbon Development Plan
LDCF	Least Developed Countries Fund
LDCs	Least Developed Countries
LGED	Local Government Engineering Department
MMR	Mixed-Method Research
MoA	Ministry of Agriculture
MoDMR	Ministry of Disaster Management and Relief
MoEFCC	Ministry of Environment Forest and Climate Change
MPEMR	Ministry of Power, Energy and Mineral Resources
MTBF	Medium-Term Budget Framework
NAP	National Adaptation Plan
NAPA	National Adaptation Program of Action
NCC, B	Network on Climate Change, Bangladesh

NDA	National Designated Authority
NDCs	National Determined Contributions
NGOAB	NGO Affairs Bureau
NGOs	Non-governmental Organisations
PD	Project Director
PKSF	Palli Karma-Sahayak Foundation
SCCF	Special Climate Change Fund
SDGs	Sustainable Development Goals
SIDS	Small Island Developing States
TIB	Transparency International Bangladesh
UNDP	United Nations Development Programme
UNESCAP	United Nations Economic and Social Commission for Asia and the Pacific
UNFCC	United Nations Framework Convention on Climate Change
USAID	United States Agency for International Development
USD	United States Dollar
WAI	Weighted Average Index
WARPO	Water Resources Planning Organisation
WB	World Bank

List of Figures

Fig. 2.1 Climate risk governance framework. *Source* Edwards et al. (2017, p. 83) 13
Fig. 2.2 Conceptual governance framework for climate change adaptation and disaster risk reduction integration. *Source* Forino et al. (2015, p. 381) 14
Fig. 2.3 Links between sustainable development, climate change and policies in these areas. *Source* Swart et al. (2003, p. S21) 17
Fig. 2.4 Tree diagram indicating the path for net high climate change effects in developing countries. *Source* Prabhakar (2010) 18
Fig. 3.1 Flowchart of the triangulation process in data collection, analysis and interpretation. *Source* Modified from Uddin (2019) 39
Fig. 3.2 Administrative units of Bangladesh. *Source* Bangladesh National Portal (2024) 41
Fig. 3.3 Map of Noakhali district showing the study area location. *Source* Banglapedia (2023) 43
Fig. 3.4 Map of Hatiya Upazila area. *Source* Authors' fieldwork (2019) 44
Fig. 3.5 Farmers packing crops in the study area. *Source* Authors' field survey (2019) 45
Fig. 3.6 Cyclone shelter/school in the study area. *Source* Authors' fieldwork (2019) 48
Fig. 3.7 Map of the Meghna Estuary. *Source* Hossain and Momtaz (2024) 49
Fig. 3.8 Research design flow chart 55
Fig. 3.9 Interview data analysis process using NVivo12. *Source* Modified from Kabir (2012) 58

Fig. 4.1 Organisational framework of climate change-related institutions and policies in Bangladesh. *Source* Modified from Ayers et al. (2014), Pervin (2013), and Huq and Rabbani (2011) 67
Fig. 4.2 Bangladesh Climate Change Trust fund governance structure. *Source* Rai et al. (2014) 70
Fig. 4.3 Schematic presentation of institutional structure. *Source* Ministry of Environment, Forest and Climate Change (MoEFCC, 2009) 80
Fig. 5.1 Map of non-governmental organisations operating in 19 coastal districts. *Source* Program Development Office for Integrated Coastal Zone Development Plan (2005) 109
Fig. 5.2 Institutional framework of the Christian Commission for Development in Bangladesh. *Source* Hossain and Momtaz (2024) 114
Fig. 5.3 Christian Commission for Development in Bangladesh project implementation framework. *Source* Christian Commission for Development in Bangladesh (2017) 124
Fig. 5.4 Dwip Unnayan Songstha climate change-related decision-making procedures. *Source* Hossain and Momtaz (2024) 126
Fig. 5.5 Non-governmental organisation funding sources. *Source* Adopted from Froelich (1999) and Viravaidya and Hayssen (2009) 128
Fig. 5.6 Radio Sagar Dwip's logo. *Source* http://dusbangladesh.org 135
Fig. 6.1 Community (respondent) age structure. *Source* Fieldwork (2019) 146
Fig. 6.2 Respondent education level. *Source* Fieldwork (2019) 147
Fig. 6.3 Respondent occupation. *Source* Fieldwork (2019) 148
Fig. 6.4 Typical fishing boats and fishers in Hatiya. *Source* Fieldwork (2019) 149
Fig. 6.5 Respondent house types. *Source* Fieldwork (2019) 150
Fig. 6.6 Typical kutcha house in Hatiya. *Source* Fieldwork (2019) 150
Fig. 6.7 Perceptions of climate change by respondents. *Source* Fieldwork (2019) 151
Fig. 6.8 Perceptions and level of climate events by respondents. *Source* Fieldwork (2019) 152
Fig. 6.9 Coastal erosion in Hatiya Island. *Source* Fieldwork (2019) 152
Fig. 6.10 Responses to climate events. *Source* Fieldwork (2019) 153
Fig. 6.11 Perceptions of respondents on the effects of climatic events. *Source* Fieldwork (2019) 154
Fig. 6.12 Destruction of reserve forest due to coastal erosion in Hatiya Island. *Source* Fieldwork (2019) 155
Fig. 6.13 Overall WAI by stakeholder relationship. *Source* Fieldwork (2019) 156

Fig. 6.14 Respondents' perceptions of accountability. *Source* Fieldwork (2019) 158
Fig. 6.15 Respondents' perceptions of transparency. *Source* Fieldwork (2019) 160
Fig. 6.16 Respondents' perceptions of resource availability. *Source* Fieldwork (2019) 162
Fig. 6.17 Climate change governance status at the local level by overall WAI 164
Fig. 6.18 Damaged village roads in the study area. *Source* Fieldwork (2019) 169
Fig. 6.19 Illegal logging in the reserve forest of Hatiya. *Source* Fieldwork (2019) 170

List of Tables

Table 2.1 Climate institutions and activities in Bangladesh 15
Table 2.2 Status of climate change governance in developing countries 19
Table 2.3 Criteria for evaluating good governance 24
Table 2.4 Principal components of climate change governance criteria ... 26
Table 3.1 Upazilas and unions of Noakhali district 42
Table 3.2 Details of the study villages 45
Table 3.3 Deadliest tropical cyclones hitting the Bangladesh coast (Noakhali-Chattogram region) 46
Table 3.4 Key informants 53
Table 3.5 Perception index levels 56
Table 4.1 Major climate change cell activities focusing on climate change 69
Table 4.2 Fund allocation by the Bangladesh Climate Change Trust 71
Table 4.3 Bangladesh Climate Change Resilience Fund investment project disbursement 72
Table 4.4 Policies, programs and plans relating to climate change in Bangladesh 75
Table 4.5 Outlined activities in Eighth Five Year Plan relating to climate change 77
Table 4.6 Participant responses to climate change governance challenges 86
Table 5.1 Details of selected non-governmental organisations 110
Table 5.2 Roles of surveyed non-governmental organisations in climate change-related issues 111
Table 5.3 Work experience of surveyed non-governmental organisations 112
Table 5.4 Christian Commission for Development in Bangladesh climate change project monitoring systems 117
Table 5.5 Selected non-governmental organisation irregularities from the Bangladesh Climate Change Trust Fund 120

Table 5.6 NGOs' performance in climate change governance practices . . . 130
Table 5.7 Successes of surveyed non-governmental organisations in climate change-related activities . . . 133
Table 5.8 Dwip Unnayan Songstha activities relating to climate change adaptation . . . 135
Table 5.9 Challenges of surveyed non-governmental organisations in climate change-related activities . . . 137
Table 6.1 Respondents' stakeholder relationship ratings . . . 155
Table 6.2 Stakeholder relationships in climate change governance . . . 156
Table 6.3 Perceptions of respondents' on accountability in climate change-related activities . . . 158
Table 6.4 Perceived views of locals on government effectiveness in climate change-related activities . . . 161
Table 6.5 Perceived views of locals on participation in climate change-related activities . . . 163
Table 6.6 Climate change governance status by WAI and OA results . . . 165
Table 6.7 Effectiveness of climate change governance as perceived by respondents . . . 166
Table 6.8 Challenges for local climate change governance . . . 168

Chapter 1
Introduction and Background

1.1 Introduction

Climate change is often considered the single biggest challenge the world is facing today (Bhuiyan, 2015; Huntjens & Zhang, 2016). It effects all countries, particularly the global south[1] (Bulkeley, 2010). It is projected that by 2050 and beyond, 150 million people globally will have become environmental refugees because of coastal flooding, river bank erosion, drought and agricultural downturn (Olsson et al., 2014). Climate change is a multifaceted global problem, and an effective response requires the involvement of numerous stakeholders and institutions (Regmi & Bhadari, 2012). A coordinated approach is necessary at multiple government levels, encompassing the spheres of politics, economics and society (Schreurs, 2010).

Governance is the practice of making and implementing decisions, rules and regulations, and collective actions within a political boundary (Sarkar, 2012; United Nations, 2012). The World Bank (2017) defines governance as 'the process through which state and non-state actors interact to design and implement policies within a given set of formal and informal rules that shape and are shaped by power' (p. 3). Good governance is necessary for a country's effective and sustained economic growth and human development. Without good governance, a country will face difficulties reaching its targeted goals. It ensures that the pledges and participation of various stakeholders in planning and implementing development works are all maintained. Good governance creates institutional environments and processes through which governmental organisations (GOs), non-governmental organisations (NGOs) and community-based organisations (CBOs) work collectively to determine a country's economic and environmental issues (Deolalikar & Jha, 2015).

Climate change governance systems incorporate cooperative and coordinated approaches from multiple stakeholders, involving formal and informal actors and

[1] The global south is defined as the economically challenged regions of the world such as Latin America, Asia, Africa and Oceania (Dados & Connell, 2012).

Md. F. Hossain and S. Momtaz, *Evaluating Climate Change Governance in Bangladesh*,
https://doi.org/10.1007/978-3-031-98486-0_1

agents (Funnell, 2001). Fröhlich and Knieling (2013) describe climate change governance as constituting a wide range of activities of coordination that expedite climate change adaptation and mitigation efforts. Climate change governance involves various processes relating to climate change plans and programs, and their effective and successful implementation for climate-vulnerable people. It also encompasses interrelationships between various stakeholders and societal coordination processes.

Bangladesh is one of the most disaster-prone countries in the world (Ministry of Environment Forest and Climate Change [MoEFCC], 2009), owing to its unique geographic location, the dominance of floodplains, low elevation, high population density, extensive poverty and the low capacity of people to withstand shock (Huntjens & Zhang, 2016; Islam, 2010, 2012; Pervin, 2013; Rai et al., 2014). Bangladesh is facing unprecedented challenges because of the adverse effects of climate change (Jaeger & Michaelowa, 2015; Mahmud & Prowse, 2012). The Intergovernmental Panel on Climate Change (IPCC) has predicted that by 2050 and beyond, climate change (in association with sea level rises) will be responsible for the submergence of 17–20% of Bangladesh's landmass (Islam, 2010; Moniruzzaman, 2019). This will create risk for 27 million people in Bangladesh by 2050 (Olsson et al., 2014). Moreover, by 2080, the World Bank (2000) has projected that 43 million people will be affected by climate change in Bangladesh, despite the country's own contributions to climate change being negligible. Bangladesh contributes less than 0.1% of global greenhouse gas emissions, whereas the United States emits 24% (Huq, 2001). The Climate Risk Index 2017 ranked Bangladesh, along with Myanmar and Pakistan, among the ten most vulnerable countries significantly affected by weather-related damage such as floods, storms and heat waves (Kreft et al., 2017).

Bangladesh's coastal areas are highly vulnerable to various natural hazards (Asaduzzaman, 2015; Brammer, 2014; Momtaz & Shameem, 2015). The country's prevailing climate governance system is a complex relationship of various stakeholders, including GOs and NGOs. They operate at different levels with regard to their respective influences. The present fragile governance system is limiting people's capacity to manage, and adjust to, climate-based shocks and stresses (Alam et al., 2011). Bangladesh's government has formulated numerous policies, plans and programs to reduce the vulnerability of affected people (Pervin, 2013). As part of its response to climate change, the Bangladesh government has established various adaptation measures over the past two decades. As a result, the number of fatalities during disasters has reduced significantly (Alam et al., 2011). Different government departments, NGOs and CSOs received USD 683.43 million in climate funds from development banks and United Nations (UN) agencies up to 2013. However, an absence of transparency persists with respect to project selection processes and funds channelled by these agencies (Transparency International Bangladesh [TIB], 2013a, 2013b).

Along with government departments, NGOs are working to develop and implement climate change plans; their participation is considered essential (Alam et al., 2011). They also support building community resilience to fight climate change effects. Moreover, they play an important role in mobilising and engaging communities, providing technical expertise and ensuring financial transparency (O'Donnell et al., 2013). For example, CARE-Bangladesh, a major NGO, has implemented the country's first climate change adaptation project, focusing on community-based adaptation in 2002. The project was 'Reducing Vulnerability to Climate Change' (Rai et al., 2014). The NGO role is often seen as one of expressing public concern over various issues, including proper policy adaptation and implementation, and corruption (Hermansen et al., 2017). Despite this, many NGOs work on climate change-related projects without the required expertise. They often emphasise the number of projects they are involved in, rather than the outcomes of those projects. They usually receive funds because of political affiliations or alleged irregularities. These irregularities indicate a poor governance system and create barriers to overcoming Bangladesh's many climate change-related problems (Bhuiyan, 2015; Islam, 2012). Above all, the lack of coordination in various governmental departments, and competition among GOs and NGOs often creates challenging situations where climate-vulnerable people cannot access the expected support.

1.2 Study Context and Significance

Various studies have reported on the climate change governance scenarios of developing countries (Ayers et al., 2014; Hermwille, 2011; Huq et al., 2011; Prabhakar, 2010). The studies' findings narrate the status of climate change governance in these countries. Although some similarities exist in the problems faced by these countries, the findings do not necessarily prescribe the same solutions to the persistent problems faced by developing countries. Climate change governance systems differ from country to country. Therefore, any policy guidance regarding climate change governance enforcement and improvement should be specific to each country. Considering these issues, it is vital to evaluate Bangladesh's specific climate change governance situation.

The present literature suggests no uncertainty regarding the role that climate change governance plays in ensuring that climate change-related projects are successful (Wilson & Termeer, 2011). Hence, many developed and developing countries, such as China, Mali, Indonesia and Brazil, have developed climate change governance systems, although many face difficulties due to the system's ineffectiveness (Leon, 2016; Qi et al., 2007; Ward et al., 2012). Thus, effective climate change governance systems are still lacking in many developing countries, including Bangladesh.

In Bangladesh, a significant amount of donor financial support gives impetus to climate change-related activities, but these suffer from inherent management deficiencies (Rai et al., 2014). Owing to the lack of effective climate change governance, the outcomes expected from relevant projects are not realised (Bhuiyan, 2015). Therefore, it is important to evaluate existing climate change governance in Bangladesh closely and precisely.

Bangladesh has achieved economic growth of 7.2% in the fiscal year 2016/17 (World Bank, 2018) with an average annual GDP growth of 6.4% between 2016 and

2021. In 2023, Bangladesh's economy is expected to grow by 5.3% according to the Asian Development Bank (ADB). Climate change issues are closely linked to sustainable development. It is therefore imperative to maintain a focus on climate change governance issues to continue and enhance the country's development initiatives. It is crucial to understand the issue and explore Bangladesh's current climate change governance activities to support the country's overall development. This book ultimately proposes an improved climate change governance framework based on the study's findings.

The significance of this book can be described in different ways. First, it contributes to a deeper understanding of Bangladesh's climate change governance practices. Bangladesh is facing serious environmental challenges because of climate change, and this book will evaluate the intrinsic relationships of climate governance and environmental governance. Moreover, this book will investigate the state of, and changes in, climate governance practices in Bangladesh over the years. Further, this book will assess the growing relationship of GOs and NGOs in climate change governance. It will also identify the major successes and challenges of existing climate governance systems based on the performances of various climate change-related initiatives in Bangladesh.

This book will focus on the guidelines for climate change governance that will support decision-making in Bangladesh. The empirical data and information collected through the field survey will help explain and conceptualise the issues. This book will examine perceptions of climate governance in two particular communities in the selected study sites and will recommend an improved governance framework for climate change-vulnerable communities; one that will improve the result of GOs and NGOs from the national to the local level. This book will contribute to the literature on climate change governance. This book will be helpful to climate change victims, climate change governing departments, agencies, NGOs, academics, research organisations and policy-makers.

1.3 Study Objectives

This research aims to evaluate climate change governance practices in Bangladesh. It focuses on the climate change governance issues of selected communities at a subdistrict level, and the roles of GOs and NGOs in climate change governance in Bangladesh.

To achieve these aims, this research has established the following specific objectives:

1. to examine the nature and salient features of GOs in climate change governance in Bangladesh
2. to examine the nature and salient features of NGOs in climate change governance in Bangladesh

3. to examine the level of community perception of climate change governance in Bangladesh.

The research addresses the following questions:

1. what climate change-related plans and programs have been adopted by different stakeholders?
2. what is the present state of institutional frameworks for climate change governance in Bangladesh?
3. what are the challenges of climate change governance faced by GOs and NGOs?
4. what is the level of participation of local communities in the climate change governance system?
5. what are the changes of climate change governance practices over time?
6. what are the effective climate change governance frameworks for the GOs and NGOs?

1.4 Methodology and Approaches

This research has determined various systems and criteria of climate governance practices through a substantial review of secondary materials and literature to address the research questions. It commenced with an empirical investigation to evaluate the nature and salient features of climate governance practices of GOs and NGOs in Bangladesh. The governance criteria, used in this study to evaluate climate governance, were adopted from the literature review. After the collection of data and information, major findings and recommendations are suggested for a system of effective climate change governance.

The literature review explores various issues, such as theoretical climate change governance considerations, the nature and context of climate governance practices in Bangladesh, criteria to evaluate climate governance, stakeholders in climate change governance, climate change linkages with sustainable development, climate governance status in developing countries, and the financial matters associated with climate change issues.

This study uses a mixed-method research (MMR) approach that is considered a clear and understandable research methodology (Creswell, 2013; Ostlund et al., 2011). This approach combines both quantitative and qualitative data in the social and behavioural sciences (Graff, 2002; van der Roest et al., 2015). The key philosophical aspect of MMR approach is pragmatism. This emphasises the relationships between research questions and the expected outcomes from those questions (Cameron, 2009; Tashakkori & Teddlie, 2003). Triangulation is used to assemble different types of data that could be analysed in different ways to better interpret the research results (Creswell, 2013; Johnson et al., 2007; Onwuegbuzie & Collins, 2007; Patton, 2002).

For this study, three different stakeholder groups, including representatives of GOs, NGOs and communities, were chosen to evaluate climate governance issues in Bangladesh. Various government departments working on climate change issues were selected and interviewed to gain opinions on climate change governance practices. A few NGOs were then selected, based on their involvement in climate change-related activities, ranging from the national to the local level. NGO staffs were interviewed to obtain their views on climate governance practices in Bangladesh. To gain community perspectives, a study area was selected, considering the research objectives and questions. The lives and livelihoods of a community are directly linked with climate change issues and the perceptions of residents in the study area were documented to present the climate governance system at the local level. Study site selection was based on various issues, such as the location, stakeholders' engagement in climate change-related activities and access to research activities. Hatiya Upazila, a subdistrict located in the south-central part of Bangladesh, is well-recognised as having experienced the effects of climate change. Because of the area's climate vulnerability, climate governance practices are evolving there, alongside other disaster management issues. Apart from this, the views of national-level experts on climate change were considered, to validate the observations obtained from other key informants. A detailed description of the selection methodology used for householders and key informants, the evaluation criteria of climate governance, the research setting and study sites are discussed in Chap. 3.

1.5 Book Structure and Content

This book consists of seven chapters. A brief outline of the chapters is discussed below.

Chapter 2 reviews important literature on the climate change governance practices of GOs and NGOs in Bangladesh. The review focuses on the theoretical context of effective climate governance, stakeholders in climate governance, status of climate change governance in developing countries, criteria for evaluating climate change governance, linkages of climate change with development, and associated financial issues.

Chapter 3 focuses on the research methodologies, design and methods used in the study. It presents in-depth views of mixed-method phenomenological research, data collection techniques and analysis methods. An outline of the study area is also given in this chapter.

Chapter 4 provides the key informant interview (KII) results based on their evaluation of the climate governance practices of GOs in Bangladesh. It demonstrates the roles of GOs in planning and implementing various climate change-related plans and programs. It also focuses on the reflection of climate governance issue in various GO actions.

Chapter 5 presents the climate governance practices of the surveyed NGOs in Bangladesh. It then describes the role of NGOs in planning and implementing various climate change-related projects and programs, and their adopted strategies in addressing climate governance issues.

Chapter 6 focuses on community perceptions of climate governance issues, presenting the demographic characteristics of interviewed householders in selected villages within the study area. It explains the participation rates of local communities in climate governance practices. It ends with a discussion about the challenges of community-level climate change governance.

Chapter 7 presents the conclusion and recommendations. It includes a summary of the research's major empirical findings concerning theoretical and policy implications, along with recommendations. It also highlights some possible future research directions relating to this study.

References

Alam, K., Shamsuddoha, M., Tanner, T., Sultana, M., Huq, M. J., & Kabir, S. S. (2011). The political economy of climate resilient development planning in Bangladesh. *Institute of Development Studies Bulletin, 42*(3), 52–60.

Asaduzzaman, M. (2015). *Livelihood vulnerability of women in the context of climate change impacts: Insights from coastal Bangladesh* [Unpublished doctoral thesis], University of Newcastle, Australia.

Ayers, J., Huq, S., Wright, H., Faisal, A. M., & Hussain, S. T. (2014). Mainstreaming climate change adaptation into development in Bangladesh. *Climate and Development, 6*(4), 293–305. https://doi.org/10.1080/17565529.2014.977761

Bhuiyan, S. (2015). Adapting to climate change in Bangladesh. *South Asia Research, 35*(3), 349–367. https://doi.org/10.1177/0262728015598702

Brammer, H. (2014). *Climate change, sea-level rise and development in Bangladesh*. The University Press.

Bulkeley, H. (2010). Cities and the governing of climate change. *Annual Review of Environment and Resources, 35*(1), 229–253. https://doi.org/10.1146/annurev-environ-072809-101747

Cameron, R. (2009). A sequential mixed model research design: Design analytical and display issues. *International Journal of Multiple Research Approaches, 3*(2), 140–152.

Creswell, J. W. (2013). *Research design: Qualitative, quantitative and mixed methods approaches*. Sage.

Dados, N., & Connell, R. (2012). The global south. *Contexts, 11*, 12–13.

Deolalikar, A. B., & Jha, S. (2015). Overview. In A. B. Deolalikar, S. Jha, & P. F. Quising (Eds.), *Governance in developing Asia: Public service delivery and empowerment*. Edward Elgar Publishing.

Fröhlich, J., & Knieling, J. (2013). Conceptualising climate change governance. In J. Knieling & W. Leal Filho (Eds.), *Climate change governance* (pp. 9–26). Springer.

Funnell, D. (2001). *Institutions and governance in mountain areas, in mountain communities between subsidy, subsidiarity and sustainability*. Paper presented at the WMS 2001, University of Bern, Institute of Geography. www.wms2001.ch/cd/

Graff, J. C. (2002). Mixed method research. In A. A. Tashakkori & C. Teddlie (Eds.), *Handbook of mixed methods in social and behavioral research* (pp. 46–66). Sage.

Hermansen, E., McNeill, D., Kasa, S., & Rajão, R. (2017). Co-operation or co-optation? NGOs' roles in Norway's international climate and forest initiative. *Forests, 8*(3), 64. https://doi.org/10.3390/f8030064

Hermwille, L. (2011). *The race to low-carbon economies has started*. www.germanwatch.org/klima/lce11.htm

Huntjens, P., & Zhang, T. (2016). *Climate justice: Equitable and inclusive governance of climate change*. http://www.thehagueinstituteforglobaljustice.org/wp-content/uploads/2016/04/Climate-Justice-April-2016.pdf

Huq, S. (2001). Climate change and Bangladesh. *Science, 294*(5547), 1617. https://doi.org/10.1126/science.294.5547.1617

Huq, S., Reid, H., Konate, M., Rahman, A., Sokona, Y., & Crick, F. (2011). Mainstreaming adaptation to climate change in least developed countries. *Climate Policy, 4*(1), 25–43. https://doi.org/10.1080/14693062.2004.9685508

Islam, N. (2010). Climate change and the challenge to social policies. *Global Social Policy, 10*(1), 18–20.

Islam, M. A. (2012). Innovative approach of ecosystem-based adaptation in coastal Bangladesh. In *Ecosystem-based adaptation* (pp. 85–106).

Jaeger, M. D., & Michaelowa, K. (2015). Global climate policy and local energy politics: Is India hiding behind the poor? *Climate Policy, 16*(7), 940–951. https://doi.org/10.1080/14693062.2015.1058239

Johnson, R. B., Onwuegbuzie, A. J., & Turner, L. A. (2007). Towards a definition of mixed methods research. *Journal of Mixed Methods Research, 1*(2), 112–133.

Kreft, S., Eckstein, D., & Melchior, I. (2017). *Global climate risk index 2017*. https://germanwatch.org/de/download/16411.pdf

Leon, A. G. (2016). Climate change governance in megadiverse countries: The case of REDD+ in Latin America. *International Journal of Climate Change: Impacts and Response, 8*(3), 61–80.

Mahmud, T., & Prowse, M. (2012). Corruption in cyclone preparedness and relief efforts in coastal Bangladesh: Lessons for climate adaptation? *Global Environmental Change, 22*(4), 933–943. https://doi.org/10.1016/j.gloenvcha.2012.07.003

Ministry of Environment, Forest and Climate Change. (2009). *Bangladesh climate change strategy and action plan 2009*. MoEFCC, Government of the People's Republic of Bangladesh.

Momtaz, S., & Shameem, M. (2015). *Experiencing climate change in Bangladesh: Vulnerability and adaptation in coastal regions*. Elsevier.

Moniruzzaman, A. N. M. (2019). *How climate change is affecting livelihoods in Bangladesh* [video]. My Justice. https://justicehub.org/article/major-general-anm-muniruzzaman-climate-change-is-affecting-livelihoods-in-bangladesh/

O'Donnell, M., Rashid, A. K. M., Steele, P., Hedger, M., Lee, J., Islam, K. M. N., Khondker, R. K., Rahman, S., Sarkar, D. K., & Sinha, N. C. (2013). Bangladesh climate public expenditure and institutional review. In R. Shaw, F. Mallick, & A. Islam (Eds.), *Climate change adaptation actions in Bangladesh* (pp. 365–385). Springer.

Olsson, L., Opoondo, M., Tschakert, P., Agrawal, A., Eriksen, S. H., Ma, L., Ma, S., Perch, L., & Zakieldeen, S. A. (2014). *Livelihoods and poverty*. https://www.ipcc.ch/pdf/assessment-report/ar5/wg2/WGIIAR5-Chap13_FINAL.pdf

Onwuegbuzie, A. J., & Collins, K. M. T. (2007). A typology of mixed methods sampling designs in social science research. *Qualitative Report, 12*(2), 281–316.

Ostlund, U., Kidd, L., Wengstrom, Y., & Rowa-Dewar, N. (2011). Combining qualitative and quantitative research within mixed method research designs: A methodological review. *International Journal of Nursing Studies, 48*(3), 369–383. https://doi.org/10.1016/j.ijnurstu.2010.10.005

Patton, M. Q. (2002). *Qualitative research and evaluation methods* (3rd ed.). Sage.

Pervin, M. (2013). *Mainstreaming climate change resilience into development planning in Bangladesh*. http://pubs.iied.org/10045IIED

Prabhakar, S. V. R. K. (2010). Promoting adaptation and disaster risk reduction in the post-Kyoto climate regime. In J. J. Pereira, R. Shaw, & J. Pulhin (Eds.), *Climate change adaptation and disaster risk reduction: Issues and challenges* (pp. 77–98). Emerald.

Qi, Y., Ma, L., & Zhang, L. (2007). Climate change governance in China: A case study. *China Population, Resources and Environment, 17*(2), 8–12. https://doi.org/10.1016/S1872-583X(07)60007-2

Rai, N., Huq, S., & Huq, M. J. (2014). Climate resilient planning in Bangladesh: A review of progress and early experiences of moving from planning to implementation. *Development in Practice, 24*(4), 527–543. https://doi.org/10.1080/09614524.2014.908822

Regmi, B. R., & Bhadari, D. (2012). Climate change governance and funding dilemma in Nepal. *TMC Academic Journal, 7*(1), 40–55.

Sarkar, A. N. (2012). Green branding and eco-innovations for evolving a sustainable green marketing strategy. *Asia-Pacific Journal of Management Research and Innovation, 8*(1), 39–58. https://doi.org/10.1177/2319510x1200800106

Schreurs, M. A. (2010). Multi-level governance and global climate change in east Asia. *Asian Economic Policy Review, 5*(1), 88–105. https://doi.org/10.1111/j.1748-3131.2010.01150.x

Tashakkori, A., & Teddlie, C. (Eds.). (2003). *The handbook of mixed methods in social & behavioral research*. Sage.

Transparency International Bangladesh. (2013a). *Climate finance in Bangladesh: Governance challenges and way out (executive summary)*. https://www.ti-bangladesh.org/beta3/images/max_file/es_cfg_executive_summary_english.pdf

Transparency International Bangladesh. (2013b). *Climate finance in Bangladesh: Governance challenges and way out*. https://www.ti-bangladesh.org/beta3/index.php/en/activities/4086-tib-demands-effective-measures-to-ensure-governance-and-reduce-corruption-in-climate-finance-2

United Nations. (2012). *Report of the United Nations conference on sustainable development*. http://www.un.org/ga/search/view_doc.asp?symbol=A/CONF.216/16&Lang=E

van der Roest, J.-W., Spaaij, R., & Van Bottenburg, M. (2015). Mixed methods in emerging academic subdisciplines: The case of sport management. *Journal of Mixed Methods Research, 9*(1), 70–90. https://doi.org/10.1177/1558689813508225

Ward, P. J., Paul, W. P., van Buuren, M. W., & Marfai, M. A. (2012). Governance of flood risk management in a time of climate change: The cases of Jakarta and Rotterdam. *Environmental Politics, 22*(3), 518–536. https://doi.org/10.1080/09644016.2012.683155

Wilson, E., & Termeer, C. (2011). Governance of climate change adaptation: Introduction to the special issue. *Climate Law, 2*, 149–157. https://doi.org/10.3233/CL-2011-031

World Bank. (2000). *Bangladesh: Climate change and sustainable development*. http://documents.worldbank.org/curated/en/906951468743377163/pdf/multi0page.pdf

World Bank. (2017). *World development report 2017: Governance and the law*. http://www.worldbank.org/en/publication/wdr2017

World Bank. (2018). *World economic prospects: Broad-based upturn but for how long?* www.worldbank.org

Chapter 2
Literature Review—Climate Change Governance in Bangladesh—Perspectives of Governmental and Non-governmental Organisations

2.1 Climate Change Governance—Theoretical Considerations

In general, governance is described as the set of decisions, actors, processes, institutional structures and mechanisms, incorporating the division of authority and underlying norms involved in determining a course of action (Adger et al., 2010; Pelling, 2011). This concept can be applied to climate change governance. An important finding from this definition is that governance is not related to public-sector actors alone but encompasses all relevant actors in decision-making. These actors maintain some arrangements, often in association with each other, for new initiatives and creating partnerships (Forino et al., 2015). Governance often conveys a 'Janus face'[1] (Swyngedouw, 2005) that moves away from an ideal decision-making process; from collective goals and benefits towards the legitimisation of privileged actors with different goals and priorities (Forino et al., 2015).

2.2 Effective Climate Change Governance

Climate change represents one of the modern world's most pressing challenges. According to Pelling (2011) a governance system is 'composed of multiple actors including public, private or civil society organisations held together through formal or informal institutions that reproduce the balance of power and direction of development pathways in society' (p. 102). A strong shift towards low-carbon development is recognised as necessary for effective climate change governance (Never, 2012). Various stakeholders, including local government organisations, regulatory agencies, business representatives and NGOs highlight the need for improvement in climate

[1] A 'Janus' face has two contrasting aspects.

Md. F. Hossain and S. Momtaz, *Evaluating Climate Change Governance in Bangladesh*,
https://doi.org/10.1007/978-3-031-98486-0_2

change adaptation governance systems. In the field of mitigation governance, more emphasis is given to the response of local communities compared with reactions from national and international agencies (Wilson & Termeer, 2011). Climate change governance is a complex technical issue. To date, policy development has given more attention to governance issues and encouraged GOs and NGOs to work efficiently (Mintrom & Luetjens, 2017). Effective climate governance must focus on the interactions between public and private sectors if it is to increase the success of climate change-related programs.

Governance systems can be considered legitimate when they are transparent, accountable, fair and equitable (Termeer et al., 2011). Governance is often influenced by a country's political strategies. It demands the attention of various social forces important to communities and their cultural values (McCarney et al., 2011). Good governance is a prerequisite to climate change policy success and funding. Effective climate change governance, as an indispensable part of a country's governance, makes it possible for a country to reduce its vulnerability to climate change. Effective climate change governance focuses on the challenges of developing policies and implementing various programs and tasks as efficiently as possible. It brings significant changes to climate change-related activities through its dynamic character and makes people more resilient to climate change-induced disasters.

A climate risk governance framework represents various controllable and uncontrollable characteristics (see Fig. 2.1). The modifiers of climate change risk focus on controlled features (e.g. exposure and vulnerability) rather than uncontrolled hazards. Effective climate change governance maintains a bottom-up approach and empowers people. It provides an opportunity to access quality information, and integrates climate change effects into future socioeconomic development planning. Thus, the resilience of livelihoods and infrastructure improves, and the capability of the poor to cope with changing climatic condition increases (African Development Bank, Asian Development Bank, Department for International Development, European Commission, & Federal Ministry for Economic Cooperation and Development—Germany, 2003). Effective climate change governance also develops a platform for the non-government sector to work efficiently (McCarney et al., 2011). It operates as a major framework (Gero et al., 2011) that enables integration with broader climate change governance of institutions and adaptation processes (Boyd & Ghosh, 2013).

Effective climate change governance is often a significant challenge at national, regional and local levels in unstable political systems and where problematic bureaucratic infrastructure exists (Barua et al., 2014a, 2014b). Patt (2017) notes that the main obstacles to effective climate change governance are not financial issues; rather, they are associated with growing knowledge, networks and institutions. Effective climate change governance is an unavoidable phenomenon that ensures development goals in the global south (Boyd & Ghosh, 2013). Effective climate governance works at multiple levels. Multi-level governance supports complex relations between the actors associated with it (Marquardt, 2017). When there is a lack of effective climate governance, a country's plans and programs usually fail to tackle the future climate change risk. Bangladesh is in a difficult situation to respond to climate change-induced disasters because of a general lack of effective climate governance.

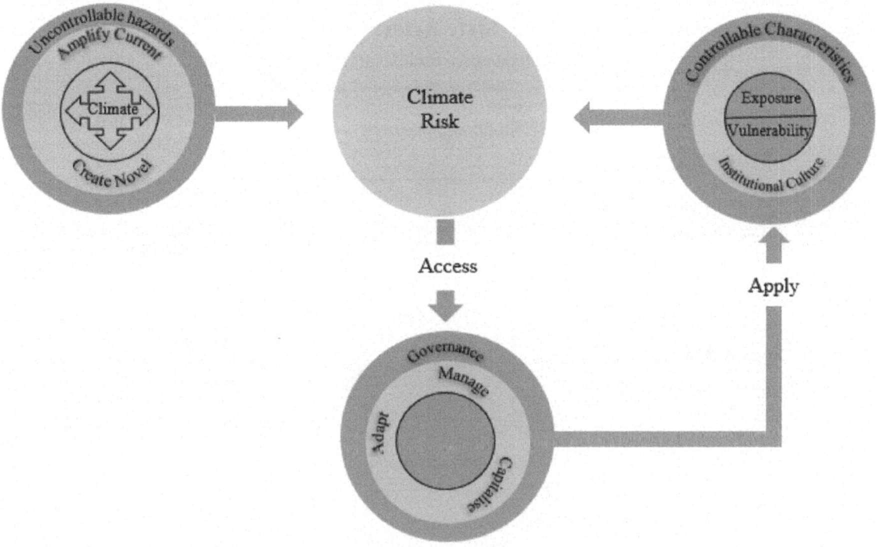

Fig. 2.1 Climate risk governance framework. *Source* Edwards et al. (2017, p. 83)

2.3 Climate Change Governance Stakeholders

Climate change governance can reduce the effects of climate-induced disasters and related risks (Forino et al., 2015). Climate change governance is a diverse and complex issue that incorporates a number of state, market and social actors, and requires a relationship be developed among them (Forino et al., 2015). Climate change issues require integrated actions at multiple levels of governance systems (Rabe, 2007; Schreurs, 2010). Blair (2017) argues that governance processes are mainly coordinated by governments, but that inputs from various relevant stakeholders must be incorporated, particularly at the local level where results are most often realised. Various actors and their interrelationships operate within a conceptual framework for climate change adaptation (CCA) and disaster risk reduction (DRR) integration (see Fig. 2.2). This framework comprises different arrangements of state, market and social actors, and encompasses the relationships between these actors. National, regional and local governments have specific roles and responsibilities in CCA and related mitigation plans and programs (Schreurs, 2010).

Various actors develop their own modes of operation in climate change-related activities; these are interlinked and engaged in coordinated and cooperative processes (Kooiman & Jentoft, 2009; Rhodes, 1996). Termeer et al. (2011) define climate change governance as collaborations between public and private actors that meet collective goals. Private sector actors and citizens can promote improved governance processes by participating in public decision-making (Munaretto & Klostermann, 2011). Different stakeholders, both individuals and organisations engaged in climate change governance are often challenged by prevailing chaotic conditions. It

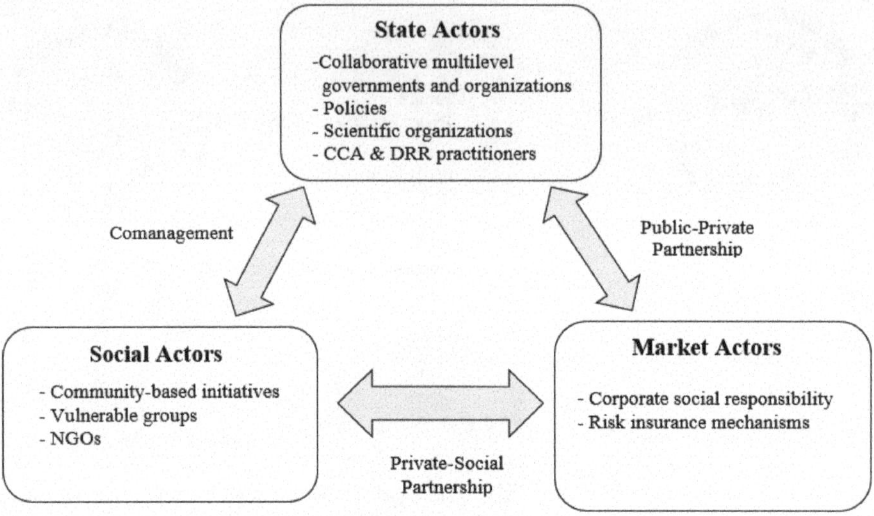

Fig. 2.2 Conceptual governance framework for climate change adaptation and disaster risk reduction integration. *Source* Forino et al. (2015, p. 381)

is therefore imperative to maintain an effective involvement in overall development and adaptation practices (Biesbroek et al., 2011).

Organisations work actively to develop climate change-related knowledge and actions (Francesch-Huidobro, 2012). Governments play a major role in conveying the findings of a climate change governance system to the people in any country. The government of Bangladesh (GoB) invested over USD 10 billion to tackle climate change and reduce the effects of natural disasters from 1980 to 2010 (GoB, 2018). Nevertheless, it is difficult for the government to rescue all from the adverse effects of climate change (Rahman & Huang, 2019). As such, government organisations precede with climate change-related initiatives in cooperation with relevant stakeholders such as donors, NGOs, civil society organisations (CSOs) and the community (see Table 2.1). Non-state actors, without having direct involvement in the process of governing, also play an important role in ensuring that national governments comply with international agreements, such as the United Nations Framework Convention on Climate Change (UNFCC) and Kyoto Protocol (Bulkeley, 2010). Local governance approaches address climate change issues through encouragement, awareness, institutions and capacity development (Bulkeley & Kern, 2006).

Climate vulnerability and resilience each have a major role in developing adaptive capacity and risk perception at the community level (Fatti & Patel, 2013). Bangladesh has implemented a significant number of community-level CCA projects over the years. Many of these have been supported by local governments, NGOs and community members (Rahman & Huang, 2019). Community-based knowledge on climate trends and adaptive strategies can be essential where information on climate change

Table 2.1 Climate institutions and activities in Bangladesh

Groups of institutions	Major areas of activities	Institutions (examples)
Government	Research and modelling	Bangladesh Agricultural Research Council, Bangladesh Agricultural Research Institute, Bangladesh Rice Research Institute, Bangladesh Institute of Development Studies, Institute of Water and Flood Management (IWFM), universities (Bangladesh University of Engineering and Technology, Dhaka University, Bangladesh Agricultural University)
	Program planning/ implementation of adaptation strategies	Ministry of Environment, Forest and Climate Change, Ministry of Food and Disaster Management, Department of Environment, Bangladesh Climate Change Trust, Ministry of Planning, Ministry of Finance, Ministry of Water Resources, Ministry of Agriculture, Ministry of Health, Local Government Engineering Department, Water Resources Planning Organisations, Ministry of Social Welfare
	Immediate disaster response	Directorate of Relief and Rehabilitation, Disaster Management Bureau, armed forces, coastal guard, district administration, Red Crescent Society, Cyclone Preparedness Program
	Climate data generation	IWFM, Centre for Environmental and Geographic Information Services, Bangladesh Meteorological Department, Bangladesh Water Development Board
Local government	Emergency measures, Pre- and post-disaster, relief and rehabilitation, food security	Union Parishad, Upazila Parishad, Municipal Corporation
Non-governmental organisation	Motivation, promotion, awareness raising, training	Bangladesh Rural Advancement Committee, Bangladesh Centre for Advanced Studies, Centre for Participatory Research and Development, Network on Climate Change Bangladesh, Christian Commission for Development in Bangladesh, Centre for Natural Resources Studies, COAST Trust, Disaster Forum, ActionAid Bangladesh
Community	Activities related to disaster risk reduction and climate change adaptation for during, pre- and post-disaster situations	Road-raising, homestead-raising, protective works (e.g. embankment building and repairing), cropping pattern, culture of climate tolerant cropping, species, crop diversification

Source Modified from Government of Bangladesh (GoB) (2012, pp. 64–65)

effects is lacking (Ayers et al., 2014). Community participation in local environmental policy and planning processes results in greater accountability and effectiveness (Abels, 2007). Community members, as with other actors (such as government and NGOs) are indispensable; they provide significant support for the governance networks, and improve climate adaptation outcomes (Keskitalo & Kulyasova, 2016).

NGOs encompass various roles similar to those of businesses, governments and charities, but they differ widely because of their size, sector focus, religious alignment, functions, relationships with partner organisations, organisational arrangements, and other factors (Gauri & Galef, 2005). NGOs play an important role in successful CCA and DRR integration (IPCC, 2012). NGOs also mediate between different actors and provide substantial professional, institutional and financial support to local governance systems (Hayes & Persha, 2010). NGOs, with the cooperation of communities, work actively to meet the needs of national- and local-level climate policy (IPCC, 2007).

Various actors and groups are involved in different phases of climate resilience development planning in Bangladesh (Alam et al., 2011). An effective climate change governance system ensures inclusion and participation of various relevant stakeholders.

2.4 Climate Change and Sustainable Development

Climate change is linked strongly to the sustainability of global socio-ecological systems. Sustainable development meets basic human needs and ensures long-term ecological sustainability (Holden et al., 2014). Sustainable development focuses on more than economic growth (UN, 1987). It is linked to climate change issues, and emphasises relevant policy formation, decision-making and implementation processes (Adger et al., 2003; Daniell et al., 2010). The climate change issue is however positioned in the development arena as part of environmental concerns within the wider aspect of sustainable development. Effective climate change governance increasingly contributes significantly in several ways to the overall development processes occurring throughout the world. Good governance acts as an essential ingredient for both rapid growth and sustainable economic development in a country (Kochanek, 2000). Climate change-related risks are providing opportunities for both empowering and institutionalising sustainable development (Adger et al., 2003).

Climate change and sustainable development have been described in largely separate circles in both research and policy (Swart et al., 2003; Vijaya Venkata Raman et al., 2012). However, this issue has now become so important that it attracts the attention of the broader community outside the research and policy areas. A sustainable development approach often addresses current social, economic and environmental problems at multiple levels (Swart et al., 2003). IPCC (2001) has noted that adaptation can be a cost-efficient technique to diminish the climate risks of a project or development, while helping significantly to fulfil the objectives of sustainable

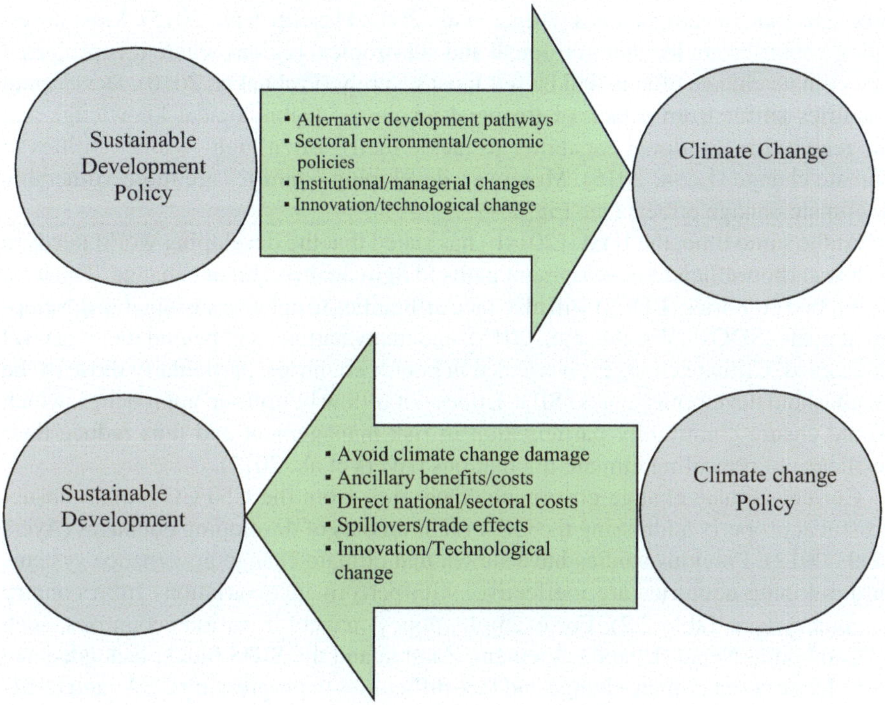

Fig. 2.3 Links between sustainable development, climate change and policies in these areas. *Source* Swart et al. (2003, p. S21)

development. The motto 'climate-smart development starts at home' (Meadowcroft, 2009) is certainly apt here.

Climate change issues can be a serious threat to sustainable development in many developing countries, including Bangladesh (Pervin & Moin, 2014). Figure 2.3 demonstrates how sustainable development policy is addressing climate change and how climate change policies are working towards sustainable development. Sustainable development and climate change issues are linked to both enhance development efforts and avoid adverse climate change-related impacts.

2.5 Climate Change Governance Status in Developing Countries

Climate change is an issue that cuts across many scientific and technical fields, creating numerous significant and persistent problems (Sarkar, 2012). The effects of climate change have become a major problem for people in developing countries (IPCC, 2014a). Developing countries are more vulnerable to climate change problems than developed ones (Adger et al., 2003; Haselip et al., 2015). Most developing countries are located in tropical and sub-tropical regions where it is projected that climate change effects will be felt most severely (Prabhakar, 2010). Developing countries suffer from a lack of financial resources, technological knowledge and the required institutional capability to tackle their—often high—vulnerabilities to climate change (Leon, 2016). Moreover, developing countries are more vulnerable to climate change effects (see Fig. 2.4).

At the same time, the IPCC (2014b) has stated that the developing world needs to follow carbon-efficient development paths to fight against climate change. The least developed countries (LDCs) will also face difficulties to achieve sustainable development goals (SDGs) (Wright et al., 2015), and many nations will be find their survival threatened. Climate change governance in poorer countries, particularly those of the small island developing states (SIDS), does not follow bottom-up approaches, which would ensure community participation in risk management and thus reduce their resilience to prevailing climate fluctuations (Ayers et al., 2012).

Current climate change governance measures, from the UNFCC remain unsuccessful at properly addressing the CCA requirements of developing countries (Ayers et al., 2012). Previous studies have shown that climate change governance systems in developing countries are ineffective, with performance variations from country to country (see Table 2.2). For example, policy-makers in various countries, such as Cambodia, Nepal, Uganda, Vietnam, Zambia and the SIDS block, lack adequate knowledge about climate change and face difficulties in progressing CCA- and DRR-related programs (Christoplos et al., 2016; Pulhin et al., 2010; Robinson & Dornan, 2016; Scobie, 2016).

Mitigation is an anthropogenic intervention to reduce the sources or enhance sinks of greenhouse gases (IPCC, 2001). Research posits mitigation as vital for developed countries and it covers the highest responsibility for climate change. In contrast, adaptation is an adjustment of natural and human systems in response to actual or expected climatic stimuli or their effects (IPCC, 2007). Adaptation is practised more

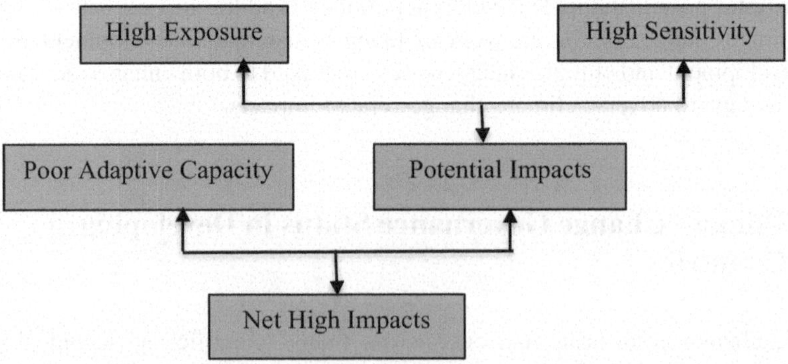

Fig. 2.4 Tree diagram indicating the path for net high climate change effects in developing countries. *Source* Prabhakar (2010)

Table 2.2 Status of climate change governance in developing countries

Authors	Country/region	Major challenges in climate change governance
Qi et al. (2007)	China	Lack of national capacity to support climate policy-making; structural problems due to decentralised government; lack of required leadership to deal with problems, inadequate capacities of local government; limited institutional frameworks
Schreurs (2010)	East Asia	Lack of sufficient financial capacity or technological knowledge; inappropriate policy for local levels; improper policy guidelines; deficiency of coordination among government- and non-governmental organisations
Kumar Sharma (2011)	Nepal	Lack of proper policy and planning; weakness in institutional arrangements; lack of meaningful participation of community-level actors; less focus on local knowledge
Huq et al. (2011)	LDCs	Lack of institutional, economic and financial capacity to cope with climate change effects; dependency on external aid; financial constraints; limited resources; lack of well-defined roles and responsibilities
Ward et al. (2012)	Indonesia	Weak administrative and regulatory frameworks; lack of multi-level, multi-scale and multi-actor governance; inadequate collaboration, cooperation and transparency among stakeholders; poor performance in planning across policy sectors
Luque et al. (2012)	Ecuador	Institutional weakness; financial constraints; lack of experts and technical capacity; limited public participation
Transparency Maldives (2015)	Maldives	Lack of access to required information on climate change projects; absence of transparency in decision-making in climate finance governance; weak institutional arrangements
Haselip et al. (2015)	Developing countries[2]	Limited capacity for reduction of greenhouse gas emissions; inadequate technology transfer in the context of climate change; poor organisational and institutional capacities

(continued)

[2] Developing countries are those having limited industrial and economic activity and where people generally have low incomes (Cambridge Dictionary, 1999). Moreover, these countries have low standard of living, inequality, poor health, and inadequate education systems (Global Economy, 2020).

Table 2.2 (continued)

Authors	Country/region	Major challenges in climate change governance
Chu et al. (2015) and Jaeger and Michaelowa (2015)	India	Absence of dynamic multi-sector governance institutions; inadequate civil society actors to highlight climate adaptation planning and implementation; local government dependency on participatory processes to expedite climate adaptation efforts; poor performance of political leadership
Dany (2015) and Pulhin et al. (2010)	Cambodia	Lack of local experts, poor institutional capacity, low performance of climate change committees, dependency on donor funds, minimal community participation
Kabiri (2016)	Thailand	Lack of effective public participation; problematic institutional frameworks; bureaucratic reluctance to share governance with non-state actors; inadequate constitutional declaration
Robinson and Dornan (2016) and Scobie (2016)	Small Island Developing States	Poor quality governance in national adaptation projects; inadequate flow of adaptation funds; high administration fees; domestic constraints; lack of required human and technical resources, political will and accountability in institutional governance frameworks
Leon (2016)	Latin America	Weak mitigation and adaptation policy; lack of information on climate change financial flows; widespread corruption; political priorities of the donors and the recipient countries
Christoplos et al. (2016)	Nepal, Uganda, Vietnam and Zambia	Emerging climate change policies; lack of administrative and legal arrangements; limited number of professional staff; lack of political will; highly donor dependent; poor public involvement
Mariano Gutiérrez et al. (2017)	Peru and Chile	Inadequate knowledge and capacity of public and private organisations; lack of required information and resources; limited community participation; weakness in decision-making
Adenle et al. (2017)	Africa	Limited institutional capacity; deficiency of finance; technology limitation; poor physical infrastructure; political instability; lack of public awareness

in the global south, as mitigation-adopting capacities are lower here, and vulnerability is greater (Ayers & Huq, 2009). National, regional and local governments in developing countries have standard and complementary roles in advancing climate change mitigation and adaptation strategies. They are responsible for successful policy, planning and program development and implementation. Based on these studies, most developing countries speak out strongly in international climate change negotiations, taking the lead in adaptation planning. In contrast, Haselip et al. (2015) focused on mitigation approaches as the focus for a few developing countries' (see Table 2.2).

Despite the presence of particular vulnerabilities and resource scarcities in developing countries, top-down approaches can be effective in dealing with governance challenge (Scobie, 2017). Several developing countries have been using strategic plans to fight climate change since 2007 (Hermwille, 2011). Developing countries are also making significant pledges regarding climate change mitigation. For instance, the Maldives and Costa Rica have taken the challenge to become carbon neutral by 2020 and 2021, respectively (Hermwille, 2011). After reviewing 19 journal articles and reports (see Table 2.2) it is evident that various issues relating to climate change governance are common to many developing countries. These include: (1) a lack of institutional arrangements (policy, administration and capacity); (2) poor performance of implementing agencies, and (3) a lack of funds and resources. Other issues for developing countries include political commitment, bureaucratic culture, leadership quality, public awareness and the participation levels of different stakeholders.

2.6 Climate Change Governance Status in Bangladesh

Bangladesh is engaged in different climate-related activities—such as policy development, budget allocation, fund collection and project implementation—that are similar to many other climate change-affected countries. Bangladesh is in a vulnerable condition for climate change due not only to nature's unpredictability, but also to a lack of effective governance (Bhuiyan, 2015). Bangladesh, like many other Asian countries, is facing institutional and governance constraints in CCA and mitigation (Bhuiyan, 2015). It therefore becomes difficult to promote the meaningful participation of all stakeholders at various levels of climate change governance. Different government departments, NGOs and donor agencies are implementing specific programs and action plans to reduce the effects of climate change-induced hazards in Bangladesh (Pulhin et al., 2010). As a result, the number of fatalities due to various disasters has reduced significantly over the last two decades (Alam et al., 2011).

Policy-makers operate at various levels and at different points concerning climate change in Bangladesh. GOs focus on different aspects to tackle climate change-related challenges at the public and private levels. Government is a key player in identifying and rectifying prevailing barriers to climate change governance. Over the last 20 years, GOs and NGOs have helped to develop several policies and action plans. These are commonly called 'foundation documents' for addressing climate change mitigation and adaptation. These policies and plans allocate substantial funds and have implemented development programs for CCA and mitigation. In Bangladesh, the mainstreaming[3] of climate change issues has evolved through various processes, such as project planning and implementation, information sharing, capacity building, resource mobilisation, and political commitment (Ayers et al., 2014).

Suffering due to climate change will be exacerbated by the existence of ineffective climate governance systems in Bangladesh (Chowdhury, 2012). The MoEFCC has overall responsibility for dealing with climate change issues. It has a limited number of skilled personnel and resources for the effective implementation of various climate projects (Scholte, 2014). Corruption reduces the effects of many activities relating to climate change effects (Bhuiyan, 2015). Government agencies involved in the allocation and use of climate finance for relevant projects are often subject to political interference or the dictates of government officials. The views of political parties on how to deal with climate issues also play an important role in the implementation of climate change-related projects. On a positive note, the Bangladesh government has developed a national climate fund from its own resources, and made positive moves towards improving climate finance at the national and local levels.

TIB has highlighted various problems in meeting climate project and program goals throughout Bangladesh; these create countless extra problems at the local level. There is an absence of vulnerability mapping and priority setting, inadequate participation of local communities in planning and monitoring systems, low quality of project work due to political interference, and a lack of meaningful compliance systems. Owing to inadequate funding, many important initiatives recommended in the Low-Carbon Development Plan (LCDP) have not yet begun in Bangladesh (Hermwille, 2011). Climate change has also not been given priority in the country's planning processes. Hence, the enormous amount of funds that could be used to address the effects of climate change are unlikely to result in any significant outcomes (Scholte, 2014). Government agencies do not conduct the necessary consultations with affected communities when preparing programs, projects and documents (TIB, 2013). These are complex to implement and require skilled coordination, and the cooperation of various stakeholders.

High levels of political commitment exist for proper management of climate change issues, but government departments, NGOs and ordinary people are not sufficiently aware of these (Bahauddin et al., 2016). Because of the lack of effective governance in Bangladesh, limited opportunities exist to manage climate change issues efficiently. As a result, powerful external groups, such as development aid donors, can influence the present system significantly (Pelling, 2011). Bangladesh has established various long-term mitigation measures, whereas adaptation and resilience efforts tend to be relatively immediate and short term. This disjointed approach may result in poor synergies in the future (Fisher, 2013). Bangladesh has undertaken several initiatives to implement programs focusing on climate change issues that strengthen capacity in all sectors; and the consciousness of different stakeholders regarding climate change issues is increasing daily.

---

[3] In the climate change context, mainstreaming refers to the incorporation of climate change issues into various plans, policies, programs and strategies, rather than undertaking adaptation and mitigation measures separately (Ayers et al., 2014).

2.7 Climate Change Governance Evaluation Criteria

Several well-recognised criteria for evaluating governance are posited by different authors and organisations (see Table 2.3). Governance can be described as an international and internal issue where values of fairness, equality, integrity and accountability are the most accepted universal social contracts (World Economic Forum, 2017). The World Governance Indicators project has developed six dimensions of governance: voice of accountability, political stability and lack of violence, governance effectiveness, regulatory quality, rule of law, and corruption control (Kaufmann et al., 2011). Effective governance is based on various conditions, such as participation, accountability, the rule of law, transparency, political stability and the absence of violence, equity and inclusiveness, corruption control, effectiveness and efficiency, and resources. Policy coordination, institutional capacity, awareness building, problem solving, integrity, and consensus are also used to assess governance effectiveness (Barua et al., 2014a, 2014b; Cadman et al., 2016; Kaufmann et al., 2011; UNESCAP, 2009; World Economic Forum, 2017). The successful adaptation and implementation of various projects relating to climate change depends upon a country's capacity to deal with climate governance issues.

This study will focus on five major criteria to evaluate the climate change governance system of Bangladesh (see Table 2.3): accountability, transparency, participation, government effectiveness and resources. These criteria were considered the most useful to evaluate climate change governance practices in Bangladesh, because these are considered the major criteria (as revealed in the literature review). These criteria also represent contextual issues that can be compared with other criteria more effectively.

Accountability is a major criterion of good governance for both actors and institutions, and indicates the ultimate performance of governance mechanisms (Biermann et al., 2010; Scobie, 2017). Various actors, organisations and leaders are accountable to the relevant authority for their activities, following some sort of standard which is termed 'accountability' (Mees & Driessen, 2018). Accountability determines which institutions are responsible for particular tasks, defines clearly specific roles and responsibilities, and identifies how these institutions perform successfully. GO and NGO accountability is an inevitable issue for ensuring effective climate governance (Disaster Forum, 2013). Accountability maintains relationships among actors and also ensures responsiveness (Mees & Driessen, 2018).

Transparency is a major mechanism for securing accountability and legitimacy in climate governance systems (Biermann et al., 2010; Gupta & Mason, 2016). Transparency is demonstrated through the extent to which actions and information are disclosed to stakeholders. It covers various aspects such as access to information, timeliness, the effort required to reach affected and vulnerable communities, and comprehensibility. Transparency and accountability are essential to engage and empower people fighting against climate change effects, deliver effective feedback, and create collaborative effort for better results (Transparency-Initiative, 2010).

Table 2.3 Criteria for evaluating good governance

Sources/ criteria	Kaufmann et al. (2011)	Barua et al. (2014a, 2014b)	Cadman et al. (2016)	World Economic Forum (2017)	UNESCAP (2009)
	Voice and accountability	Clarity of role and responsibility	Accountability	Fairness	Participation
	Political stability and absence of violence/ terrorism	Institutional capacity	Transparency	Equality	Rule of law
	Government effectiveness	Policy coordination	Inclusiveness	Integrity	Transparency
	Regulatory quality	Transparency	Resources	Accountability	Responsiveness
	Rule of law	Stakeholder engagement	Equality		Consensus-oriented
	Control of corruption		Democracy		Equity and inclusiveness
			Agreement		Effectiveness and efficiency
			Dispute settlement		Accountability
			Behavioural change		
			Problem solving		
			Durability		

Government effectiveness represents the institutional capacity for policy formulation and implementation, the quality of public services, the degree of independence from political interventions, managing resources, addressing technical issues, and the level of government commitment to policy (Barua et al., 2014a, 2014b; Kaufmann et al., 2010, 2011). Government effectiveness is often seen as a result-based approach of good governance in dealing with climate change issues (Balsiger & Van Deveer, 2012; Biermann et al., 2010).

Climate change governance encompasses participatory approaches, along with local perspectives, various stakeholders, and community representatives (Sprain, 2016). Stakeholder participation in planning and implementation processes makes climate change-related programs more effective (Barua et al., 2014a, 2014b). As climate change governance is a highly complex field, relevant stakeholder participation can address more effectively, resulting in better outcomes (Kabiri, 2016). Participation expedites results from climate governance at the local, national, regional and international levels (Jodoin et al., 2015).

Resources are important to implement climate-related plans and programs, and are necessary for both climate adaptation and mitigation (Allred et al., 2016). Resources, as a criteria, relates to the sources of funding, the processes of using funds and the effectiveness of that use. A country can prompt climate change-related activities only if adequate resources are available (Adenle et al., 2017). Bangladesh is mobilising climate funds from different sources for various projects and programs to address adverse climatic events that may take place in future.

Table 2.4 presents the principal components based on the study conducted, including subcomponents and sources.

2.8 Climate Change Finance

Climate change finance is an important aspect of the successful implementation of mitigation and adaptation plans and programs in various sectors. The UNFCC focused on climate finance issues at the world climate summit held in Germany in 2017 (Cui & Huang, 2018). Globally, developing countries face challenges related to fulfilling LCDPs, because of the channelling of limited funds to work for climate change problems (Adenle et al., 2017). Efficient financing is a prerequisite for an empowered governance system in developing countries. Funding problems can present significant obstacles to the enhancement processes of climate change policy elements (Kern & Alber, 2009). A well-organised, permanent and adequate financial mechanism can improve social and economic imbalances, so that a country becomes more resilient to climate change's adverse conditions (McCarney et al., 2011).

Developed countries are providing increasing amounts of financial support to developing countries, to help them tackle the adverse effects of climate change. Countries that are more vulnerable to climate change are more likely to receive greater financial aid (Weiler et al., 2018). Most funds are delivered by developed countries to their more vulnerable counterparts as grants or loans. Aid allocation for CCA based on various criteria. According to the recipient merit model, donors prefer that participants have 'good' institutions and policies[4] (Chu et al., 2015; Weiler et al., 2018). Thus, good governance and good policies are important in the allocation of climate change funds. Developed countries have committed to providing US $100 billion per year from 2020 to fulfil the needs of developing countries (UNFCC, 2009). The global south does not consider climate finance as a charity on behalf of developed countries; rather, it is considered compensation (Chowdhury, 2012).

The GoB has developed its own climate funds for improving various sectors affected by climate change. The GoB allocated US $100 million when it established BCCSAP from 2009 to 2010 (Chowdhury, 2012). The BCCTF collected US $347 million from national revenue between 2010 and 2014 (Pervin & Moin, 2014). Bangladesh also received funds through BCCRF, mainly from development

[4] These demonstrate the recipient's value of democratic institutions, political stability, control of corruption, rule of law and government effectiveness.

Table 2.4 Principal components of climate change governance criteria

Principal components	Subcomponents (indicators)	Sources
Accountability	Institutional responsibility Specific roles and responsibility Performances of the institutions Democratic practices in decision-making	Bannerman et al. (2011), Kaufmann et al. (2011), Mees and Driessen (2018) and Mir and Bala (2014)
Transparency	Ease of access to information Comprehensiveness Timeliness Comprehensibility Financial transparency Level of effort to reach the affected and vulnerable communities	Baroi and Panday (2018), Barua et al. (2014a, 2014b), Gupta and Mason (2016), Hahn et al. (2015) and Lopa and Ahmad (2016)
Participation	Application of participatory approaches in decision-making Stakeholder participation at the local level Stakeholder participation at the national level	Barua et al. (2014a, 2014b), Huitema et al. (2016), Kabiri (2016) and Lopa and Ahmad (2016)
Government effectiveness	Institutional capacities for policy formulation Implementation quality of public services The degree of independence from political interventions Addressing technical issues Level of government commitments to various policies Manage resources	Hayes and Persha (2010), Kaufmann et al. (2011), United Nations Economic and Social Commission for Asia and the Pacific (2009) and Weiler et al. (2018)

(continued)

Table 2.4 (continued)

Principal components	Subcomponents (indicators)	Sources
Resources	Sources of resources Disbursement of financial resources The processes of utilisation The effectiveness	Adenle et al. (2017), Barua et al. (2014a, 2014b), Mir and Bala (2014) and Viravaidya and Hayssen (2001)

partners. These funds are channelled through both national and international agencies. Bangladesh has also faced governance challenges regarding climate finance. As Bangladesh is receiving climate funds from various sources, the transparency, accountability and integrity relating to spending these funds has become a matter of concern. How finance will be managed is a major issue for climate change projects in Bangladesh. The Bangladesh Climate Change Trust (BCCT) has experienced varying degrees of corruption, including climate funds being withdrawn without implementing any works, or undertaking only partial work in around 50 projects (Khan, 2015).

Owing to the lack of required funding, the Bangladesh government faces difficulties when implementing the projects and programs recommended in BCCSAP (GoB, 2015). Bangladesh expects compensation in the form of grants—not loans—for climate change mitigation and adaptation. Loans would be yet another burden for the climate-affected people of Bangladesh. TIB (2013) called on the government to develop capacity and improve negotiation skills when dealing with partner countries and bodies to secure grants rather than loans for addressing climate change, especially in relation to adaptation. The Seventh Five-Year Plan (2016–2021) also emphasised fund use and ensuring the involvement of community voices for meaningful citizen participation in future climate change challenges (GoB, 2015).

2.9 Climate Change Governance Evolution in Bangladesh

Climate change governance practices in Bangladesh have a varied historical background. Before the country's independence (Bangladesh was previously East Pakistan) from West Pakistan (presently Pakistan), a Water Pollution Control Ordinance of 1970 (Government of East Pakistan, 1970) existed. After independence was gained 1971, Bangladesh started to develop policies and plans to manage the environment, as well as climate change issues. The Water Pollution Control Ordinance of 1970 was cancelled and replaced by the Environment Pollution Control Ordinance, 1977 (GoB, 1977). There were no specific national policies relating to climate change in the past. The GoB, through the MoEFCC, has adopted policies and institutional initiatives to cope with the effects of climate change. Various policies such as the

National Environment Policy 1992, the Forest Policy 1995, the New Agriculture Extension policy 1996, the National Wetland Policy 1998, the National Water Policy 1999, the Landuse Policy 2001, the Renewable Energy Policy 2008, the Agriculture Policy 2009, and the National Health Policy 2010 were all focused on climate change issues (Alam et al., 2013). The Bangladesh Environment Conservation Act of 1995 and the Water Management Plan 2001 both focused on future environmental problems arising from climate change. Climate change adaptations are the focus of national and sectoral plans, and strategies to enhance climate resilience in the system overall (Nafee et al., 2020).

The MoEFCC is responsible for various tasks, such as preparing national communication, formulating adaptation programs, making approvals under clean development mechanism projects, leading international negotiations and mainstreaming climate change at sectoral levels. The MoEFCC has made significant progress in developing some policies and institutional arrangements to minimise the effects of climate change. These include the development of a National Adaptation Programme of Action (NAPA), the Bangladesh Climate Change Strategy and Action Plan (BCCSAP), establishing a Climate Change Unit (CCU) under the MoEFCC, and a Climate Change Trust Fund (CCTF). The MoEFCC is working on climate change and adaptation with allied ministries and CSOs to implement projects and programs under NAPA and BCCSAP at the national level.

Prior to the formulation of BCCSAP in 2009, the GoB developed a number of national and sectoral strategies and action plans emphasising environmental and climate change issues. These included: (a) the National Environment Management Action Plan 1995; (b) the National Water Management Plan 2001; and (c) the National Biodiversity Strategy and Action Plan 2004 (updated in 2016). Over the last 15 years, climate change issues have been central to national affairs, and noteworthy changes have been observed in institutional settings. Various plans and policies, government bodies, academic institutions, NGOs, networks and campaigns have been developed over this period. The government established the Climate Change Cell (CCC) in 2004. This body is located in the Department of Environment (DoE), under MoEFCC with financial support from the Department for International Development (DFID). The CCC is responsible for the strengthening, coordination and management of projects on climate change mitigation and adaptation funded by the government and donors.

Bangladesh was one of the first countries to develop NAPA in 2005, in part to meet the requirements of the UNFCC. This document focused on 15 urgent priorities linked with CCA. The NAPA outlined the adverse effects of climate change and plotted a way forward for sustainable development. The 'Climate Change Trust Fund Act 2010' was approved by cabinet to tackle disasters resulting from climate change (Huq & Rabbani, 2011). The Bangladesh Climate Change Trust Fund (BCCTF) was developed by the government from its own resources, and the multi-donor-supported Bangladesh Climate Change Resilience Fund (BCCRF) was established in 2010 to address climate change. The GoB developed the BCCSAP in 2008, and updated it in 2009. The BCCSAP is the latest policy document of the government to address both mitigation and adaptation for the next 10 years. GOs and NGOs are pursuing different adaptation projects directed by BCCSAP (Rai et al., 2014). The BCCSAP focuses on six themes (GoB, 2009):

1. Food security, social protection and health
2. Comprehensive disaster management
3. Infrastructure development
4. Research and knowledge management
5. Mitigation and low-carbon development
6. Capacity building and institutional development by short and medium actions.

Climate change issues have been mainstreamed into policy-making in Bangladesh (Ayers et al., 2014). Climate change-related activities are coordinated by the National Environment Committee under the MoEFCC. This committee includes representatives from all relevant ministries and government agencies. The MoEFCC has undertaken a Comprehensive Disaster Management Program (CDMP), which encompasses a holistic approach to DRR focusing on climate change.

The government has formulated a number of high-level committees to deal with climate change issues, such as inter-ministerial committees, MoEFCC and an all-party parliamentary committee to oversee the planning and implementation processes of CCA activities in the country. Bangladesh is a pioneer country among the LDCs,[5] which have established and maintained National Funding Entities since 2010; these countries are addressing climate change. Bangladesh is the first country among the LDCs with national low-carbon resilience plans and strategies. Bangladesh was selected as a leader of the LDCs in 2005 and 2006, and is still leading these countries to seek international support for addressing climate change issues. The country is also playing a significant role in international climate change negotiations and contributes to the Conference of the Parties (COP). Climate change policy, focusing on adaptation, is a focal point for development (GoB, 2005). As a sign of this change in focus, the GoB renamed the Ministry of Environment and Forest (MoEF) the MoEFCC on 14 April 2018. This represents an endorsement of climate change's importance for the government.

Moreover, donors, NGOs and CSOs are working on specific programs and action plans to address climate change-induced hazards in Bangladesh (Pulhin et al., 2010). NGO participation in climate change-related activities can ensure that vulnerable communities are represented (Alam et al., 2013). A number of NGOs are working in different ways in climate change-related activities, such as the Bangladesh Rural Advancement Committee (BRAC), the Bangladesh Centre for Advanced Studies (BCAS), the Centre for Participatory Research and Development (CPRD), the Network on Climate Change, Bangladesh (NCCB), the Christian Commission for Development in Bangladesh (CCDB), the Centre for Natural Resources Studies (CNRS), COAST Trust, the Disaster Forum (DF), and ActionAid Bangladesh (AAB). These are actively involved at the local, national and international levels in climate

[5] Least developed countries (LDCs) are low-income countries facing severe structural impediments to attain sustainable development. These countries are so vulnerable to economic and environmental shocks and their human resources are low level. There are 47 countries in the LDC category (UN, 2020).

change research, capacity building, policy or advocacy, modelling, monitoring and evaluation, and community-level demonstrations of climate change-related projects.

This literature review has presented the significance of different aspects of climate change governance and the role of different stakeholders to ensure its greater effectiveness. Based on the literature review, the present research has presented various issues relating to climate change governance by GOs and NGOs in Bangladesh. It has also illustrated community perceptions of climate change governance to display the evolution of climate governance practices at the local level, as local communities are the ultimate beneficiaries of effective climate governance. It also mentioned some counterfactual arguments and analysis concerning climate governance to develop a result-based evaluation of the research. Among the plethora of research on climate governance, this research has focused on the role of different stakeholders, and how improving their involvement will ensure more effective outcomes in developing countries like Bangladesh. This research also addressed various issues, such as contributing to the existing knowledge and making links with the existing literature, demonstrating local consequences and providing recommendations for an effective climate governance system in Bangladesh that will bring it up to expected standards.

Chapter 3 outlines the research methodology, including the study area settings that have been undertaken to find relevant results.

References

Abels, G. (2007). Citizen involvement in public policy-making: Does it improve democratic legitimacy and accountability? The case of PTA. *Interdisciplinary Information Sciences, 13*(1), 103–116.

Adenle, A. A., Manning, D. T., & Arbiol, J. (2017). Mitigating climate change in Africa: Barriers to financing low-carbon development. *World Development, 100*, 123–132. https://doi.org/10.1016/j.worlddev.2017.07.033

Adger, W. N., Huq, S., Brown, K., Conway, D., & Hulme, M. (2003). Adaptation to climate change in the developing world. *Progress in Development Studies, 3*(3), 179–195.

Adger, W. N., Lorenzoni, I., & O'Brien. (Eds.). (2010). *Adapting to climate change: Thresholds, values, governance*. Cambridge University Press.

Alam, K., Shamsuddoha, M., Tanner, T., Sultana, M., Huq, M. J., & Kabir, S. S. (2011). The political economy of climate resilient development planning in Bangladesh. *Institute of Development Studies Bulletin, 42*(3), 52–60.

Alam, K., Tanner, T., Shamsuddoha, M., Rashid, A. K. M. M., Sultana, M., Huq, M. J., Kabir, S. S., & Ullah, S. (2013). Planning 'exceptionalism'? Political economy of climate change resilient development in Bangladesh. In R. Shaw, F. Mallick, & A. Islam (Eds.), *Climate change adaptation in Bangladesh* (pp. 387–417). Springer.

Allred, S., Schneider, R., & Reeder, J. (2016). The role of natural resource professionals in addressing climate change. *Climate, 4*(38), 1–17. https://doi.org/10.3390/cli4030038

Ayers, J. M., & Huq, S. (2009). The value of linking mitigation and adaptation: A case study of Bangladesh. *Environ Management, 43*(5), 753–764. https://doi.org/10.1007/s00267-008-9223-2

Ayers, J., Alam, M., & Huq, S. (2012). Global adaptation governance beyond 2012: Developing-country perspectives. In F. Biermann, P. Pattberg, & F. Zelli (Eds.), *Global climate governance beyond 2012*. Cambridge University Press.

Ayers, J., Huq, S., Wright, H., Faisal, A. M., & Hussain, S. T. (2014). Mainstreaming climate change adaptation into development in Bangladesh. *Climate and Development, 6*(4), 293–305. https://doi.org/10.1080/17565529.2014.977761

Bahauddin, K. M., Rahman, N., & Hasnine, M. T. (2016). Environmental reviews and case studies: Public perception, knowledge, and participation in climate change adaptation governance in the coastal region of Bangladesh using the social ecological inventory (SEI) tool. *Environmental Practice, 18*(1), 32–43. https://doi.org/10.1017/s1466046615000393

Balsiger, J., & Van Deveer, S. D. (2012). Navigating regional environmental governance introduction. *Global Environmental Politics, 12*(3), 1–17.

Bannerman, M., Rashid, M. H. O., & Rejve, K. (2011). *NGO-government partnerships for disaster preparedness in Bangladesh.* https://odihpn.org/magazine/ngo%c2%96government-partnerships-for-disaster-preparedness-in-bangladesh/

Baroi, H. S., & Panday, P. K. (2018). NGOs in local governance in Bangladesh. *South Asian Survey, 22*(2), 156–170. https://doi.org/10.1177/0971523117744149

Barua, P., Fransen, T., & Wood, D. (2014a). *Climate policy implementation tracking framework.* Retrieved from World Resources Institute: https://www.wri.org/sites/default/files/climate_tracking_framework_working.pdf

Barua, P., Fransen, T., & Wood, D. (2014b). *Climate policy implementation tracking framework.* http://wri.org/publication/climate-policy-tracking

Bhuiyan, S. (2015). Adapting to climate change in Bangladesh. *South Asia Research, 35*(3), 349–367. https://doi.org/10.1177/0262728015598702

Biermann, F., Betsill, M. M., Gupta, J., Kanie, N., Lebel, L., Liverman, D., Schroeder, H., Siebenhüner, B., & Zondervan, R. (2010). Earth system governance: A research framework. *International Environmental Agreements: Politics, Law and Economics, 10*(4), 277–298.

Biesbroek, R., Klostermann, J., Termeer, C., & Kabat, P. (2011). Barriers to climate change adaptation in the Netherlands. *Climate Law, 2*, 181–199. https://doi.org/10.3233/CL-2011-033

Blair, A. A. C. (2017). *Climate change threats and adaptation responses in small island developing states: A comparative analysis of Antigua & Barbuda and Vanuatu* [Unpublished doctoral thesis], University of Newcastle, Australia.

Boyd, E., & Ghosh, A. (2013). Innovations for enabling urban climate governance: Evidence from Mumbai. *Environment and Planning C: Government and Policy, 31*(5), 926–945. https://doi.org/10.1068/c12172

Bulkeley, H. (2010). Climate policy and governance: An editorial essay. *Wiley Interdisciplinary Reviews: Climate Change, 1*(3), 311–313. https://doi.org/10.1002/wcc.1

Bulkeley, H., & Kern, K. (2006). Local government and the governing of climate change in Germany and the UK. *Urban Studies, 43*(12), 2237–2259.

Cadman, T., Maraseni, T., Breakey, H., López-Casero, F., & Ma, H. (2016). Governance values in the climate change regime: Stakeholder perceptions of REDD+ legitimacy at the national level. *Forests, 7*(10), 212. https://doi.org/10.3390/f7100212

Cambridge Dictionary. (1999). *Cambridge Dictionary*. Cambridge University Press.

Chowdhury, F. A. (2012). *Climate change finance and governance: Bangladesh perspectives.* https://www.academia.edu/3581112

Christoplos, I., Aben, C., Bashaasha, B., Dhungana, H., Friis-Hansen, E., Funder, M., Nguyen, T.-T.-H., Khatri, D. B., Lindegaard, L. S., Mweemba, C., Le Duc, N., Nyambe, I., Ojha, H., Okiror, J., Pain, A., & Len, T.-H.-S. (2016). *Understanding sub-national climate governance: Finding from Nepal, Uganda, Vietnam and Zambia.* http://pure.diis.dk/ws/files/614369/DIIS_RP_2016_5_WEB.pdf

Chu, E., Anguelovski, I., & Carmin, J. (2015). Inclusive approaches to urban climate adaptation planning and implementation in the Global South. *Climate Policy, 16*(3), 372–392. https://doi.org/10.1080/14693062.2015.1019822

Cui, L., & Huang, Y. (2018). Exploring the schemes for green climate fund financing: International lessons. *World Development, 101*, 173–187. https://doi.org/10.1016/j.worlddev.2017.08.009

Daniell, K. A., Máñez Costa, M. A., Ferrand, N., Kingsborough, A. B., Coad, P., & Ribarova, I. S. (2010). Aiding multi-level decision-making processes for climate change mitigation and adaptation. *Regional Environmental Change, 11*(2), 243–258. https://doi.org/10.1007/s10113-010-0162-0

Dany, V. (2015). *Climate change adaptation planning in Cambodia and potential for improvements* [Ph.D.], Bond University, Australia. http://epublications.bond.edu.au/cgi/viewcontent.cgi?article=1184&context=theses

Disaster Forum. (2013). *Governance and disaster management*. https://www.disasterforum.org/index.html

Edwards, I., Burton, D., & Baker-Jones, M. (2017). Governance and climate change risk. *Governance Directions*, March, 82–87.

Fatti, C. E., & Patel, Z. (2013). Perceptions and responses to urban flood risk: Implications for climate governance in the south. *Applied Geography, 36*, 13–22. https://doi.org/10.1016/j.apgeog.2012.06.011

Fisher, S. (2013). *Low carbon resilient development in least developed countries*. http://pubs.iied.org/17177IIED/

Forino, G., von Meding, J., & Brewer, G. J. (2015). A conceptual governance framework for climate change adaptation and disaster risk reduction integration. *International Journal of Disaster Risk Science, 6*(4), 372–384. https://doi.org/10.1007/s13753-015-0076-z

Francesch-Huidobro, M. (2012). Institutional deficit and lack of legitimacy: The challenges of climate change governance in Hong Kong. *Environmental Politics, 21*(5), 791–810. https://doi.org/10.1080/09644016.2012.686221

Gauri, V., & Galef, J. (2005). NGOs in Bangladesh: Activities, resources, and governance. *World Development, 33*(12), 2045–2065. https://doi.org/10.1016/j.worlddev.2005.05.009

Gero, A., Méheux, K., & Dominey-Howes, D. (2011). Integrating community based disaster risk reduction and climate change adaptation: Examples from the Pacific. *Natural Hazards and Earth System Sciences, 11*(1), 101–113. https://doi.org/10.5194/nhess-11-101-2011

Global Economy. (2020). *Common characteristics of developing countries* [video]. https://www.youtube.com/watch?v=zXVBoRaZDFo

GoB. (2018). *Climate financing for sustainable development: Budget report 2018–19*. Retrieved from Dhaka, Bangladesh: www.mof.gov.bd

Government of Bangladesh. (1977). *The environment pollution control ordinance, 1977*. https://www.ecolex.org/details/legislation/environment-pollution-control-ordinance-1977-ordinance-no-xiii-of-1977-lex-faoc006437/

Government of Bangladesh. (2005). *National adaptation programme of action (NAPA)*. https://unfccc.int/resource/docs/napa/ban01.pdf

Government of Bangladesh. (2009). *Bangladesh climate change strategy and action plan 2009*. http://www.moef.gov.bd/site/page/97b0ae61-b74e-421b-9cae-f119f3913b5b/BCCSAP-2009

Government of Bangladesh. (2012). *Bangladesh climate public expenditure and institutional review*. Government of the People's Republic of Bangladesh.

Government of Bangladesh. (2015). *Sectoral inputs towards the formulation of the seventh five year plan (2016–2021): Climate change and disaster management*. http://www.plancomm.gov.bd/wp-content/uploads/2015/02/11a_Climate-Change-and-Disaster-Management.pdf

Government of Bangladesh. (2015). *Seventh five year plan FY2016–FY2020: Accelerating growth, empowering citizens*. https://www.unicef.org/bangladesh/sites/unicef.org.bangladesh/files/2018-10/7th_FYP_18_02_2016.pdf

Government of East Pakistan. (1970). *The East Pakistan water pollution control ordinance, 1970*. https://www.ecolex.org/details/legislation/east-pakistan-water-pollution-control-ordinance-no-v-of-1970-lex-faoc002343/

Gupta, A., & Mason, M. (2016). Disclosing or obscuring? The politics of transparency in global climate governance. *Current Opinion in Environmental Sustainability, 18*, 82–90. https://doi.org/10.1016/j.cosust.2015.11.004

Hahn, R., Reimsbach, D., & Schiemann, F. (2015). Organizations, climate change, and transparency. *Organization & Environment, 28*(1), 80–102. https://doi.org/10.1177/1086026615575542

Haselip, J., Hansen, U. E., Puig, D., Trærup, S., & Dhar, S. (2015). Governance, enabling frameworks and policies for the transfer and diffusion of low carbon and climate adaptation technologies in developing countries. *Climatic Change, 131*(3), 363–370. https://doi.org/10.1007/s10584-015-1440-0

Hayes, T., & Persha, L. (2010). Nesting local forestry initiatives: Revisiting community forest management in a REDD+ world. *Forest Policy and Economics, 12*(8), 545–553. https://doi.org/10.1016/j.forpol.2010.07.003

Hermwille, L. (2011). *The race to low-carbon economies has started*. www.germanwatch.org/klima/lce11.htm

Holden, E., Linnerud, K., & Banister, D. (2014). Sustainable development: Our common future revisited. *Global Environmental Change, 26*, 130–139. https://doi.org/10.1016/j.gloenvcha.2014.04.006

Huitema, D., Adger, W. N., Berkhout, F., Massey, E., Mazmanian, D., Munaretto, S., Plummer, R., & Termeer, C. J. A. M. (2016). The governance of adaptation: Choices, reasons, and effects. Introduction to the special feature. *Ecology and Society, 21*(3), article 37. https://doi.org/10.5751/es-08797-210337

Huq, S., & Rabbani, G. (2011). Climate change and Bangladesh: Policy and institutional development to reduce vulnerability. *Journal of Bangladesh Studies, 13*(1), 1–10.

Huq, S., Reid, H., Konate, M., Rahman, A., Sokona, Y., & Crick, F. (2011). Mainstreaming adaptation to climate change in least developed countries. *Climate Policy, 4*(1), 25–43. https://doi.org/10.1080/14693062.2004.9685508

Intergovernmental Panel on Climate Change. (2001). *Synthesis report, third assessment, working group III. Summary for policymakers*. Retrieved from Paris: http://www.ipcc.ch/ipccreports/tar/wg2/pdf/wg2TARspm.pdf

Intergovernmental Panel on Climate Change. (2007). *Climate change 2007: Mitigation. Contribution of working group III to the fourth assessment report of the Intergovernmental Panel on Climate Change*. https://www.ipcc.ch/pdf/assessment-report/ar4/wg3/ar4_wg3_full_report.pdf

Intergovernmental Panel on Climate Change. (2012). *Managing the risks of extreme events and disasters to advance climate change adaptation*. https://www.ipcc.ch/pdf/special-reports/srex/SREX_Full_Report.pdf

Intergovernmental Panel on Climate Change. (2014a). *Climate change 2014: Impacts, adaptation, and vulnerability. Summaries, frequently asked questions, and cross-chapter boxes. Contribution of working group II to the fifth assessment report of the Intergovernmental Panel on Climate Change*. A World Meteorological Organization. http://www.ipcc.ch/report/ar5/wg2/

Intergovernmental Panel on Climate Change. (2014b). *Climate change 2014: Synthesis report*. http://ar5-syr.ipcc.ch/ipcc/ipcc/resources/pdf/IPCC_SynthesisReport.pdf

Jaeger, M. D., & Michaelowa, K. (2015). Global climate policy and local energy politics: Is India hiding behind the poor? *Climate Policy, 16*(7), 940–951. https://doi.org/10.1080/14693062.2015.1058239

Jodoin, S., Ducky, S., & Lofts, K. (2015). Public participation and climate governance: An introduction. *Review of European Community & International Environmental Law, 24*(2), 117–122.

Kabiri, N. (2016). Public participation, land use and climate change governance in Thailand. *Land Use Policy, 52*, 511–517. https://doi.org/10.1016/j.landusepol.2014.12.014

Kaufmann, D., Kraay, A., & Mastruzzi, M. (2010). *The worldwide governance indicators: Methodology and analytical issues*. www.govindicators.org

Kaufmann, D., Kraay, A., & Mastruzzi, M. (2011). The worldwide governance indicators: Methodology and analytical issues. *Hague Journal on the Rule of Law, 3*(02), 220–246. https://doi.org/10.1017/s1876404511200046

Kern, K., & Alber, G. (2009). *Governing climate change in cities: Modes of urban climate governance in multi-level systems*. http://library.wur.nl/WebQuery/wurpubs/fulltext/51364

Keskitalo, E. C. H., & Kulyasova, A. A. (2016). The role of governance in community adaptation to climate change. *Polar Research, 28*(1), 60–70. https://doi.org/10.1111/j.1751-8369.2009.00097.x

Khan, M. Z. H. (2015). *Civil society's role in tackling corruption and climate finance.* www.transparency.org

Kooiman, J. A. N., & Jentoft, S. (2009). Meta-governance: Values, norms and principles, and the making of hard choices. *Public Administration, 87*(4), 818–836. https://doi.org/10.1111/j.1467-9299.2009.01780.x

Kumar Sharma, S. (2011). The political economy of climate change governance in the Himalayan region of Asia: A case study of Nepal. *Procedia—Social and Behavioral Sciences, 14*, 129–140. https://doi.org/10.1016/j.sbspro.2011.03.030

Leon, A. G. (2016). Climate change governance in megadiverse countries: The case of REDD+ in Latin America. *International Journal of Climate Change: Impacts and Response, 8*(3), 61–80.

Lopa, F. G. R., & Ahmad, M. M. (2016). Participation of CSOs/NGOs in Bangladeshi climate change policy formulation: Co-operation or co-optation? *Development in Practice, 26*(6), 781–793. https://doi.org/10.1080/09614524.2016.1200536

Luque, A., Edwards, G. A. S., & Lalande, C. (2012). The local governance of climate change: New tools to respond to old limitations in Esmeraldas, Ecuador. *Local Environment, 18*(6), 738–751. https://doi.org/10.1080/13549839.2012.716414

Mariano Gutiérrez, T., Jorge Castillo, P., Laura Naranjo, B., & Akester, M. J. (2017). Current state of goods, services and governance of the Humboldt current large marine ecosystem in the context of climate change. *Environmental Development, 22*, 175–190. https://doi.org/10.1016/j.envdev.2017.02.006

Marquardt, J. (2017). Conceptualizing power in multi-level climate governance. *Journal of Cleaner Production, 154*, 167–175. https://doi.org/10.1016/j.jclepro.2017.03.176

McCarney, P., Blanco, H., Carmin, J., & Colley, M. (2011). Cities and climate change. In W. D. S. C. S. Rosenzweig, A. Hammer, & S. Mehrotra (Eds.), *Climate change and cities: First assessment report of the urban climate change research network* (pp. 249–269). Cambridge University Press.

Meadowcroft, J. (2009). *Climate change governance.* World Bank research working policy paper no. 4941. https://papers.ssrn.com/sol3/papers.cfm?abstract_id=1407959

Mees, H., & Driessen, P. (2018). A framework for assessing the accountability of local governance arrangements for adaptation to climate change. *Journal of Environmental Planning and Management, 62*(4), 671–691. https://doi.org/10.1080/09640568.2018.1428184

Mintrom, M., & Luetjens, J. (2017). Policy entrepreneurs and problem framing: The case of climate change. *Environment and Planning C: Politics and Space, 35*(8), 1362–1377. https://doi.org/10.1177/2399654417708440

Mir, M., & Bala, S. K. (2014). NGO accountability in Bangladesh: Two contrasting cases. *VOLUNTAS: International Journal of Voluntary and Nonprofit Organizations, 26*(5), 1831–1851. https://doi.org/10.1007/s11266-014-9513-7

Munaretto, S., & Klostermann, J. E. M. (2011). Assessing adaptive capacity of institutions to climate change: A comparative case study of the Dutch Wadden Sea and the Venice Lagoon. *Climate Law, 2*, 219–250. https://doi.org/10.3233/CL-2011-035

Nafee, K. M., Real, H., & Khan, R. (2020). Climate resilient development planning in Bangladesh: A review. *Bangladesh Students' Digest on Economics and Development, 03*, 56–69.

Never, B. (2012). Who drives change? Comparing the evolution of domestic climate governance in India and South Africa. *Journal of Environment & Development, 21*(3), 362–387. https://doi.org/10.1177/1070496512449821

Patt, A. (2017). Beyond the tragedy of the commons: Reframing effective climate change governance. *Energy Research & Social Science, 34*, 1–3. https://doi.org/10.1016/j.erss.2017.05.023

Pelling, M. (2011). *Adaptation to climate change: From resilience to transformation.* Routledge.

Pervin, M., & Moin, S. M. I. (2014). *Climate finance governance in Bangladesh: Synergies in the financial landscape*. http://pubs.iied.org/pdfs/17227IIED.pdf

Prabhakar, S. V. R. K. (2010). Promoting adaptation and disaster risk reduction in the post-Kyoto climate regime. In J. J. Pereira, R. Shaw, & J. Pulhin (Eds.), *Climate change adaptation and disaster risk reduction: Issues and challenges* (pp. 77–98). Emerald.

Pulhin, J. M., Shaw, R., & Pereira, J. (2010). Climate change adaptation and disaster risk reduction: Key challenges and ways forward. In R. Shaw, J. Pulhin, & J. J. Pereira (Eds.), *Climate change adaptation and disaster risk reduction: An Asian perspective* (pp. 451–476). Emerald.

Qi, Y., Ma, L., & Zhang, L. (2007). Climate change governance in China: A case study. *China Population, Resources and Environment, 17*(2), 8–12. https://doi.org/10.1016/S1872-583X(07)60007-2

Rabe, B. G. (2007). Beyond Kyoto: Climate change policy in multilevel governance systems. *Governance: An International Journal of Policy, Administration, and Institutions, 20*(3), 423–444. https://doi.org/10.1111/j.1468-0491.2007.00365.x

Rahman, M. M., & Huang, D. (2019). Climate governance initiatives: Snapshots from Bangladesh. *Journal of Geoscience and Environment Protection, 7*(9), 131–147. https://doi.org/10.4236/gep.2019.79010

Rai, N., Huq, S., & Huq, M. J. (2014). Climate resilient planning in Bangladesh: A review of progress and early experiences of moving from planning to implementation. *Development in Practice, 24*(4), 527–543. https://doi.org/10.1080/09614524.2014.908822

Rhodes, R. A. W. (1996). The new governance: Governing without government. *Political Studies, 44*(4), 652–667. https://doi.org/10.1111/j.1467-9248.1996.tb01747.x

Robinson, S.-A., & Dornan, M. (2016). International financing for climate change adaptation in small island developing states. *Regional Environmental Change, 17*(4), 1103–1115. https://doi.org/10.1007/s10113-016-1085-1

Sarkar, A. N. (2012). Green branding and eco-innovations for evolving a sustainable green marketing strategy. *Asia-Pacific Journal of Management Research and Innovation, 8*(1), 39–58. https://doi.org/10.1177/2319510x1200800106

Scholte, M. (2014). *Climate finance governance*. https://www.giz.de/en/worldwide/14874.html

Schreurs, M. A. (2010). Multi-level governance and global climate change in East Asia. *Asian Economic Policy Review, 5*(1), 88–105. https://doi.org/10.1111/j.1748-3131.2010.01150.x

Scobie, M. (2016). Policy coherence in climate governance in Caribbean small island developing states. *Environmental Science & Policy, 58*, 16–28. https://doi.org/10.1016/j.envsci.2015.12.008

Scobie, M. (2017). Accountability in climate change governance and Caribbean SIDS. *Environment, Development and Sustainability, 20*(2), 769–787. https://doi.org/10.1007/s10668-017-9909-9

Sprain, L. (2016). Paradoxes of public participation in climate change governance. *Good Society, 25*(1), 62–80.

Swart, R., Robinson, J., & Cohen, S. (2003). Climate change and sustainable development: Expanding the options. *Climate Policy, 3*, S19–S40. https://doi.org/10.1016/j.clipol.2003.10.010

Swyngedouw, E. (2005). Governance innovation and the citizen: The Janus face of governance-beyond-the-state. *Urban Studies, 42*(11), 1991–2006. https://doi.org/10.1080/00420980500279869

Termeer, C., Dewulf, A., Rijswick, H. V., Buuren, A. V., Huitema, D., Meijerink, S., Rayner, T., & Wiering, M. (2011). The regional governance of climate adaptation: A framework for developing legitimate effective and resilient governance arrangements. *Climate Law* (2), 159–179. https://doi.org/10.3233/CL-2011-032

Transparency-Initiative. (2010). *Climate change: New frontiers in transparency and accountability*. www.transparency-initiative.org

Transparency International Bangladesh. (2013). *Climate finance in Bangladesh: Governance challenges and way out (executive summary)*. https://www.ti-bangladesh.org/beta3/images/max_file/es_cfg_executive_summary_english.pdf

Transparency Maldives. (2015). *Climate finance governance in Maldives: Pressing issues*. http://transparency.mv/files/media/2ad69df9bf7886a5063f21a7c541bd4a.pdf

United Nations. (1987). *Our common future*. http://www.un-documents.net/our-common-future.pdf
United Nations. (2020). *Least developed countries*. https://www.un.org/development/desa/dpad/least-developed-country-category.html
United Nations Economic and Social Commission for Asia and the Pacific. (2009). *What is good governance?* http://www.unescap.org/resources/what-good-governance
United Nations Framework Convention on Climate Change. (2009). *Report of the conference of the parties on its fifteenth session*. https://unfccc.int/resource/docs/2009/cop15/eng/11a01.pdf
Vijaya Venkata Raman, S., Iniyan, S., & Goic, R. (2012). A review of climate change, mitigation and adaptation. *Renewable & Sustainable Energy Reviews, 16*(1), 878–897. https://doi.org/10.1016/j.rser.2011.09.009
Viravaidya, M., & Hayssen, J. (2001). *Strategies to strengthen NGO capacity in resource mobilization through business activities*. https://www.unaids.org/sites/default/files/media_asset/jc579-strategies_ngo_en_2.pdf
Ward, P. J., Paul, W. P., van Buuren, M. W., & Marfai, M. A. (2012). Governance of flood risk management in a time of climate change: The cases of Jakarta and Rotterdam. *Environmental Politics, 22*(3), 518–536. https://doi.org/10.1080/09644016.2012.683155
Weiler, F., Klöck, C., & Dornan, M. (2018). Vulnerability, good governance, or donor interests? The allocation of aid for climate change adaptation. *World Development, 104*, 65–77. https://doi.org/10.1016/j.worlddev.2017.11.001
Wilson, E., & Termeer, C. (2011). Governance of climate change adaptation: Introduction to the special issue. *Climate Law, 2*, 149–157. https://doi.org/10.3233/CL-2011-031
World Economic Forum. (2017). *Global shapers survey*. http://www.shaperssurvey2017.org/static/data/WEF_GSC_Annual_Survey_2017.pdf
Wright, H., Huq, S., & Reeves, J. (2015). *Impacts of climate change on least developed countries: Are the SDGs possible?* http://pubs.iied.org/17298IIED/

Chapter 3
Methodology

3.1 Introduction

In the previous chapter, the theoretical framework and background on climate change governance practices on developing countries and LDCs, including Bangladesh, were established. This chapter describes the research methods used to collect and analyse the required data and information, to evaluate Bangladesh's climate change governance system. This chapter covers the following sections: research methodologies, research design, study area, data collection techniques, data analysis, validity and reliability, and ethical issues.

3.2 Research Methods

The study was conducted based on both primary and secondary data. The study area and selected cross-sectional samples of population were the sources of primary data and information. Literature such as journal articles, official published reports, books, conference papers and official websites were the main sources of secondary data and information for this study.

An exploratory approach followed by MMR was used to assess the status of climate change governance practices. The research design or plan for this study encompassed various issues. It tried to highlight existing climate change governance practices in Bangladesh from local areas, selected study sites and at the national level.

Md. F. Hossain and S. Momtaz, *Evaluating Climate Change Governance in Bangladesh*,
https://doi.org/10.1007/978-3-031-98486-0_3

3.3 Research Design

This research was conducted in accordance with social and behavioural science techniques, using both quantitative and qualitative approaches. Quantitative data can support qualitative components by recognising representative sample members, as well as different case studies. Conversely, at the design phase, qualitative data supported the quantitative research components by enhancing conceptual and instrument development. At the data collection level, quantitative data helped provide baseline information and support to avoid bias. In the data analysis stage, quantitative data helped simplify qualitative data, whereas qualitative data performed an important role by clarifying, describing, interpreting, constructing, adjusting and authenticating quantitative results (Johnson et al., 2007). The qualitative approach helped respondents reinforce their views, thoughts, opinions and clarifications.

MMR has been an acknowledged methodological approach since the early 1990s; it has been presented as a 'quiet' revolution that works to minimise tensions between qualitative and quantitative research (Graff, 2002). MMR is considered a synthesis where the researcher collects, analyses, integrates and incorporates concepts from both quantitative and qualitative data (Creswell, 2013; Johnson et al., 2007). It has emerged as a progressively influential research technique in the social and behavioural sciences (Graff, 2002; van der Roest et al., 2015). This approach is becoming clarified and connected to research practices, and is appreciated as a major research approach (Johnson et al., 2007). MMR requires the collection, analysis and interpretation of both quantitative and qualitative data, using an expansive view (Ostlund et al., 2011; Zohrabi, 2013). Moreover, this approach combines the statistical analysis of large sample cases with in-depth investigation (Creswell, 2013; Lieberman, 2005).

The key philosophical aspect of MMR is pragmatism (Cameron, 2009). Pragmatism focuses on using multiple methods within one study and emphasises the relationship between research questions and the appropriate findings from those questions (Tashakkori & Teddlie, 2003). It also focuses on pluralistic approaches to develop knowledge for greater understanding of the research issue (Creswell, 2013; Morgan, 2007). MMR has become more accepted and it provides an overarching justification by incorporating specific qualitative methods within a MMR framework (Mayoh & Onwuegbuzie, 2013). Further, the MMR paradigm provides a significant approach for developing research questions and indicating appropriate answers for those questions (Johnson et al., 2007). This research follows a MMR approach that gives an improved understanding of research problems and findings (Creswell, 2013; Ostlund et al., 2011).

However, a few limitations exist with MMR. The distinction it makes between qualitative and quantitative methods may appear simple to apply, but complexities can arise. MMR can fail to interpret real-world scenarios by linking qualitative and quantitative methods (Maghaddam et al., 2002). It is often difficult to combine different methods, and one may be more applicable than another. The findings from different methods might not address actual scenarios, and may create contradictions and confusion in the expected results. Nevertheless, this research has adopted various precautions in data collection and analysis to maintain data accuracy when presenting the results. Moreover, this study combines both qualitative and quantitative methods to obtain stronger findings.

3.3.1 Triangulation

Triangulation is the cross-checking of information and conclusions (Johnson et al., 2007). Triangulation is a strategy for improving research validity and reliability or for evaluating the results obtained from a study (Golafshani, 2003; Sufian, 1998). This was followed by research for convergence among multiple and different sources of information to form themes or classes in a study (Creswell & Miller, 2000). Triangulation is used to assemble different types of data that can then be analysed in a variety of ways to produce a better interpretation (Creswell, 2013; Onwuegbuzie & Collins, 2007). Triangulation strengthens studies by combining various methods (Patton, 2002).

In this study, triangulation incorporated different methods, including quantitative and qualitative data collection, and data analysis, as presented in Fig. 3.1. In household surveys, the validity of data and information was maintained by asking household heads questions relating to climate change governance issues in the study sites. This was conducted with a semi-structured questionnaire, then rating interviewee's opinions on a Likert scale (Davison, 2014). Further, qualitative data collection techniques (e.g. key informant interviews [KIIs] and focus group discussions [FGDs]) were followed to ensure validity, conformity and reliability of the study results.

3.4 The Study Area

This section provides an outline of the study area, detailing its geographical and socioeconomic settings. The location of the study site and its characteristics were important factors for evaluating the existing climate change governance systems of

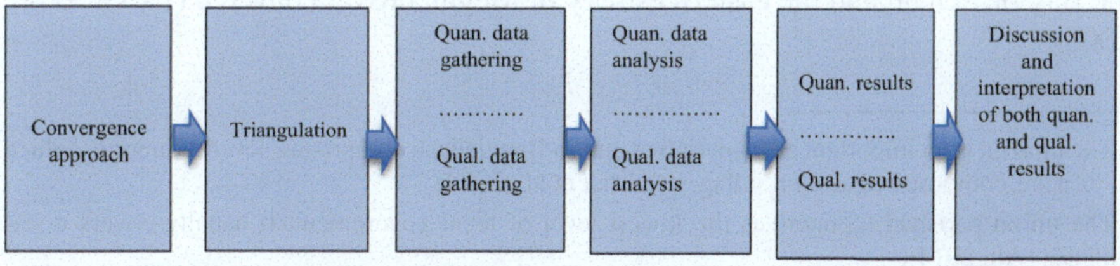

Fig. 3.1 Flowchart of the triangulation process in data collection, analysis and interpretation. *Source* Modified from Uddin (2019)

a particular area. The overall settings of the study area demonstrated how and why this area is vulnerable to climate change effects. This study site is similar to other areas on the coast that have suffered from climate change.

3.4.1 *Structure of Administrative Units in Bangladesh*

Bangladesh is a democratic republic and the prime minister is the chief executive of the government. For administration purposes, there are eight major administrative units called 'divisions' in Bangladesh. Each division is based on the major city within its jurisdiction, which acts as the headquarter of that particular division. There are 64 districts in Bangladesh and these are divided further into upazilas[1] (sub-districts) (see Fig. 3.2). These 491 upazilas work at the regional level to implement development projects from the national government. Moreover, all these divisions, districts and upazila headquarters work as urban centres in the country. The upazilas are divided into several unions,[2] with exceptions being observed in metropolitan areas. These union parishads (councils) are the smallest administrative units at the local level. These unions reach 4554 in total. They comprise multiple villages, which represent the smallest social units. The total number of villages is more than 86,000. Metropolitan areas are governed by the city corporations. At present, there are 12 city corporations and 327 municipalities in Bangladesh. The administrative structures of Bangladesh are illustrated in Fig. 3.2.

3.4.2 *Study Area Details*

The coastal area of Bangladesh is highly vulnerable to rising sea levels because of climate change effects in the Bay of Bengal, a low-lying deltaic setting (Nishat et al., 2009). The length of coastline is around 710 km and 23% of the country is a coastal zone; these areas are critically vulnerable to climate change-induced disasters (Islam, 2008). The study was conducted in two villages of the Char[3] Iswar Union of the Hatiya Upazila, in the Noakhali district under the Chattogram division. This district is located in the south-central part of Bangladesh, where the inmost part of the Bay of Bengal and the eastern estuary of Meghna River converge (Parvin et al., 2008).

[1] The upazila is an important administrative unit in Bangladesh comprising several 'unions', which in turn are composed of several villages (Nishat et al., 2009).

[2] The union parishad (council) is the lowest level of local government. It usually covers a few villages (Ali, 2014).

[3] Char is a tract of land surrounded by the waters of an ocean, sea, lake, or stream. Chars in Bangladesh are observed as 'by-product' of the rivers' hydro-morphological activities (Chowdhury & Chowdhury, 2014).

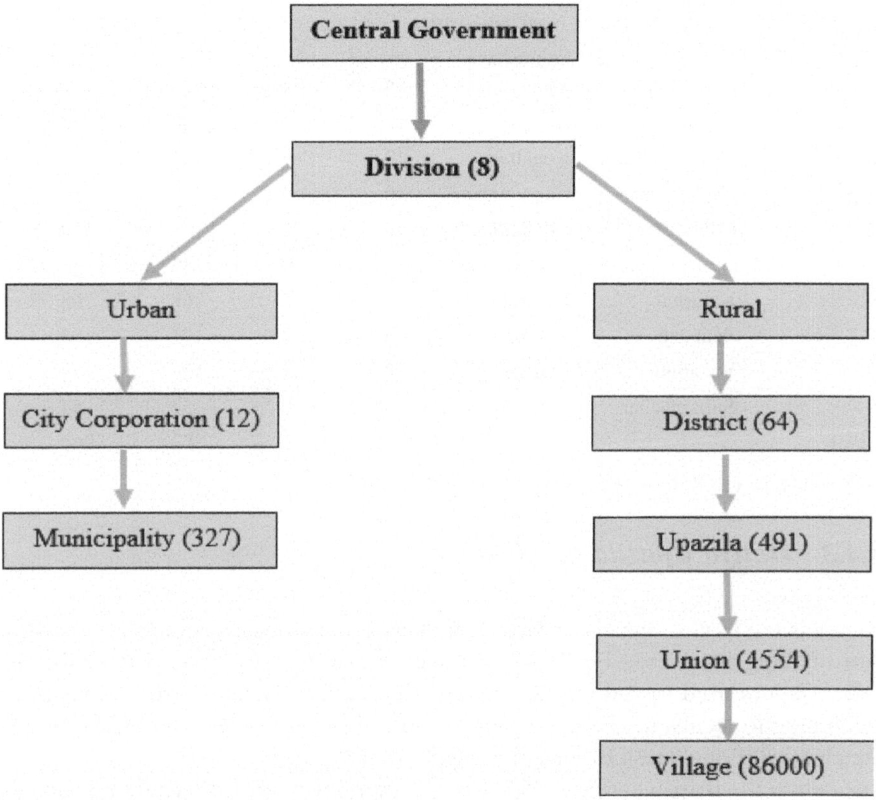

Fig. 3.2 Administrative units of Bangladesh. *Source* Bangladesh National Portal (2024)

The total area of Noakhali district is 3685.87 km^2 (1423.12 square miles), of which 1557.18 km^2 is under reserve forest. This district covers approximately 2.44% of the country's forests. Noakhali, one of the largest deltaic districts, is situated at the fringe of the Bay of Bengal. It lies between 22° 07′ and 23° 08′ north latitudes and between 90° 53′ and 91° 27′ east longitudes (see Fig. 3.1). The district is bounded on the north by Comilla district, on the east by Feni and Chattogram districts, on the south by the Bay of Bengal and on the west by Bhola and Lakshmipur districts (BBS, 2015a). The population density of Noakhali is 843 people per square kilometre. The male-to-female ratio is 101:99 (BBS, 2015a).

This district consists of nine upazilas, 91 unions, 882 mouzas,[4] and 967 villages (see Table 3.1). It also has eight municipalities with 72 wards. The upazilas of Noakhali district are Begumganj, Chatkhil, Companiganj, Hatiya, Kabirhat, Noakhali Sadar, Senbagh, Sonaimuri and Subarnachar.

[4] Mouza is the geographical expression of a unit of land area for revenue settlement and revenue collection. Since 1962, the term 'mouza' is used in settlement documents, but this is in trend now (Islam, 2015b).

Table 3.1 Upazilas and unions of Noakhali district

S. No.	Upazila	Pourashava[5]	No. of unions	No. of wards	No. of mouzas	No. of villages
1	Hatiya	Hatiya	11	9	44	62
2	Noakhali Sadar	Noakhali	13	9	166	173
3	Begumganj	Chowmuhani	16	9	178	184
4	Chatkhil	Chatkhil	10	9	108	129
5	Companiganj	Companiganj	8	9	36	45
6	Kabirhat	Kabirhat	7	9	68	68
7	Senbagh	Senbagh	9	9	97	105
8	Sonaimuri	Sonaimuri	10	9	133	148
9	Subarnachar	–	8	0	52	53
Total			91	72	882	967

Source Bangladesh Bureau of Statistics (BBS, 2015a, 2015b)

3.4.3 *Hatiya Upazila at a Glance*

Hatiya is the largest upazila of Noakhali district and is located in the south-central part of Bangladesh (see Fig. 3.3). In earlier times, this area was part of the sea. Dutch sailors used to call this area 'Hatiya', which in Dutch means 'water land'. According to locals, the name of 'Hatiya' upazila might have originated from this Dutch word (Bangladesh Bureau of Statistics [BBS], 2015a). Moreover, it is reported that during the British period (1757–1947), a high-level official travelled around the entire Hatiya Upazila on foot, as there was no alternative transport. In Bengali the word 'Hatiya' means 'walking', which led to the official declaration of the area as 'Hatiya' (Das, 2017).

The total area of Hatiya Upazila is 1508.23 km^2. The location of this upazila is 22° 07′ and 22° 35′ north latitude and 90° 56′ and 91° 11′ east longitude. It is bounded by Noakhali Sadar and Ramgati upazilas on the north, the Bay of Bengal on the south and the east, and the Manpura upazila of Bhola district on the west (BBS, 2015a). This upazila includes one municipality, nine wards, 11 unions, 44 mouzas and 62 villages. The 11 unions of this upazila include Sukhchar, Burirchar, Chandnandi, Char Ishwar, Char King, Harni, Jahajmara, Nalchira, Nijhum Dip, Sonadia and Tamaruddin. This upazila also encompass offshore islands of various sizes.

The total area of Hatiya Island is 46,682 ha (Dwip Unnayan Report, 2008). The island's topography is nearly flat. The average elevation of the island above mean sea level is approximately 2.4 m. The average length of the island in a north–south direction is approximately 48 km, and in an east–west direction is about 15 km (Kumar & Ghosh, 2012). This is an offshore island and the coastline is shifting

[5] Pourashava/Municipal Corporations areas, excluding the cities with city corporations, have municipal corporations, also known as Pourashava. In Bangladesh, pourashavas are considered the secondary towns, which hold around 40% of the urban population.

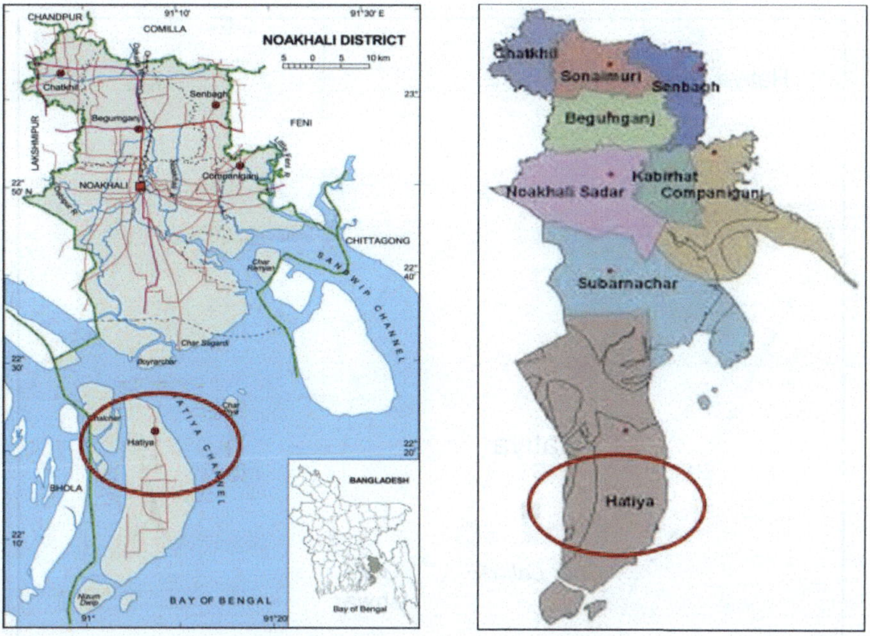

Fig. 3.3 Map of Noakhali district showing the study area location. *Source* Banglapedia (2023)

because of human activity, river discharge, tides, coastal hydrology and floods. The area is also affected by the three large river systems; namely, the Ganges (Padma), Brahmaputra (Jamuna), and the Meghna.

According to the 'Population and Housing Census 2011' report, the total population of this upazila was 452,463 (BBS, 2015b). The upazila had a population density of 300 persons per square kilometre in 2011, whereas the average population density of Bangladesh was 1015. The population growth rate in Bangladesh is 1.37% per annum, while the rate in Hatiya is three percent. The literacy rate in Bangladesh averages 73%, but the rate in Hatiya is 68%. Only five percent of Hatiya Upazila has electricity coverage, whereas the coverage throughout Bangladesh averages 70%. Around 90% of roads in the upazila are unpaved and damaged. This upazila remains behind in all national development indices (Fig. 3.4).

The people of Hatiya Upazila are mostly involved in farming and fishing. They make a significant contribution to Bangladesh's economy, especially in the fishery sector, and through honey production. The GoB declared 40 thousand acres of land here as a national park in 2001, for its natural beauty and to protect the area's biodiversity (Chowdhury, 2015). Nijhum Dwip (island), one of the most important tourist places in Bangladesh, is located on the southern side of this upazila: here, you can watch the sun rise and set from the vast, extended sea beaches on the island's southern and western sides (Fig. 3.5).

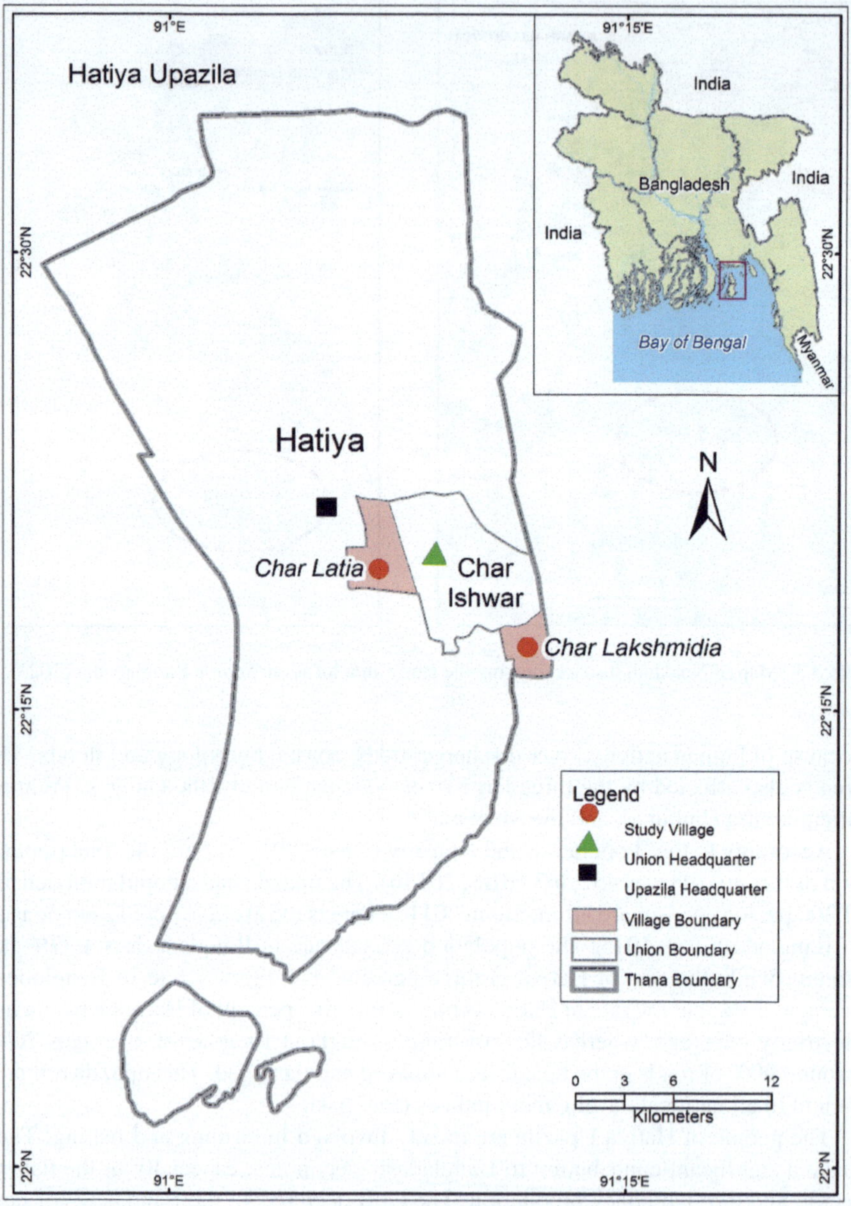

Fig. 3.4 Map of Hatiya Upazila area. *Source* Authors' fieldwork (2019)

Char Ishwar Union is located in the southern part of Hatiya Island (see Fig. 3.4). This union is located on the bank of Meghna River. The total area of this union is 26.20 km^2 and the population numbers 42,221 (Union Council, 2019). The population

Fig. 3.5 Farmers packing crops in the study area. *Source* Authors' field survey (2019)

Table 3.2 Details of the study villages

Name of district	Name of upazila	Name of union	Name of villages	Number of households
Noakhali	Hatiya	Char Ishwar	Char Latiya	577
			Char Lakshmidia	941
Total				1518

Source GoB (2015)

density of this union is 312 per square kilometre (GoB, 2015). The villages under this union number 15, and two—Char Latiya and Char Lakshmidia—were selected for this study. The number of households of Char Latiya and Char Lakshmidia is 577 and 941, respectively (see Table 3.2).

3.4.4 Disasters and Climate Events in Hatiya Upazila

The NAPA (MoEFCC, 2005) has revealed that coastal zone vulnerability in this area will be acute because of the effects of climate change. According to Quadir and Iqbal (2008), in the Bay of Bengal, 539 tropical cyclones were recorded over a 131 year period (1877–2007). A number of deadly tropical cyclones affected the Bangladesh coast over this period, especially in the Noakhali-Chattogram region, and caused significant property damage, and the loss of life (see Table 3.3).

Table 3.3 Deadliest tropical cyclones hitting the Bangladesh coast (Noakhali-Chattogram region)

S. No.	Year	Landfall date	Landfall	Wind speed (km/h)	Storm surges (m)	Human causalities	Economic loss
1	1960	31 October	Noakhali-Chattogram	193	6.1	10,000	27,783 cattle died 568,161 houses damaged 5–7 vessels capsized
2	1961	9 May	Meghna Estuary	161	3.1	11,468	25,000 cattle died
3	1963	29 May	Noakhali-Chattogram	202	6	11,520	32,617 cattle died 4787 boats and standing crops were damaged
4	1965	12 May	Barisal Noakhali	162	3.7	19,279	
5	1965	1 June	Hatiya	87			
6	1966	1 October	Noakhali	139	6.7	850	65,000 cattle died 309,000 houses damaged 1,500,000 people were badly affected
7	1970	12 November	Bhola, Meghna Estuary	224	10	300,000	Colossal damages were occurred
8	1971	6 November	Noakhali-Chattogram				
9	1974	28 November	Chattogram Cox's Bazar	161	5.1	50	1000 cattle died 2300 houses damaged
10	1975	28 November	Chattogram			83	
11	1976	21 October	Noakhali-Chattogram	80			

(continued)

Table 3.3 (continued)

S. No.	Year	Landfall date	Landfall	Wind speed (km/h)	Storm surges (m)	Human causalities	Economic loss
12	1983	15 October	Chattogram	85		43	Fishers and boats missing over 100, crops damaged 20%
13	1985	25 May	Noakhali	154	4.5–5	4264	Colossal damaged occurred
14	1991	29 April	Noakhali-Chattogram	235	7.6	133,882	Total loss was BDT 145 billion
15	1991	2 June	Noakhali-Patuakhali	83	2.5	300	Considerable number of houses and crops were damaged
16	1997	19 May	Chattogram	200	4.6	155	211,717 houses damaged 3118 cattle died

Source Alam (2014) and Quadir and Iqbal (2008)

Hatiya Upazila, because of its geographic location and poor socioeconomic conditions, is widely known as one of the most climate disaster-risk upazilas in Bangladesh (GoB, 2017). The average elevation of this upazila is lower than the mean sea level, which means that floods, tropical cyclones, storm surges, and river erosion have become common scenarios for the people of this area (Islam, 2004).

Hatiya, owing to the local conditions, has been affected by devastating cyclones and storm surges in 1960, 1961, 1970, 1985, 1991, 1994, 1997, 1998 and 2007 (Parvin et al., 2008; Union Council, 2009). Among these, the 1970, 1991 and 1994 cyclones affected the Char Iswar Union severely. These climate events caused substantial damage to settlements, livestock and crops (Islam, 2015a). It was reported that a high-level flood and storm surge occurred on the island in the 1960s; it was prolonged for 22 days, and caused colossal damage to livelihoods and properties (Das, 2017).

According to the Upazila Adminstration (2005) report, around 130,000 people have died because of cyclones and storm surges since Bangladesh's independence in 1971. In 1970, approximately 10,000 people died in a single cyclone in one night, and 80% were women and children. Another major cyclone that occurred in 1991 resulted in 6000 causalities, and over half were women (DUS, 2016) (Fig. 3.6).

Fig. 3.6 Cyclone shelter/school in the study area. *Source* Authors' fieldwork (2019)

The Meghna Estuary, a dynamic estuary, carries a major portion of flow from the Ganges–Brahmaputra River system to the Bay of Bengal (Ali et al., 2007). Owing to large water and sediment discharge, extensive erosion and accretion have taken place in this estuary. The Hatiya Upazila is located in the Meghna Estuary (see Fig. 3.7). This estuary gained a net amount of 1700 km^2 of land over the last 65 years (Sarker et al., 2011). Another study revealed that, in a 21 year (1989–2010) period, approximately 6476 ha of land were eroded from Hatiya Island, whereas around 9916 ha were accreted onto the island (Ghosh et al., 2015). Moreover, an analysis of satellite images from 1973 and 2008 illustrate that the net rate of accretion has been very high in Noakhali district, because this district is the most sediment-deposition prone in the area (Sarker et al., 2011).

Over many years, erosion has occurred in the northern part of the island, while the rate of erosion has declined slowly in the western parts. Most accretion has occurred in the island's southern and eastern parts. As a result, various areas of the island, such as Char Iswar, Char King, Tamaruddin, Sonadia, Burir Char and Jahajmara, were formed through accretion at the end of the eighteenth century, and the total land area increased (Das, 2017). However, the existence of Hatiya Island is dependent on the condition of this estuary. The people of Hatiya Island have overcome various problems over the years, but they still face trouble with cyclones, storm surges and river or coastal erosion.

An embankment that surrounded Hatiya Island to protect it from cyclone and storm surges from the Bay of Bengal was approved by the then Pakistan government in 1965. A 20–22 foot high embankment was developed by the East Pakistan Water

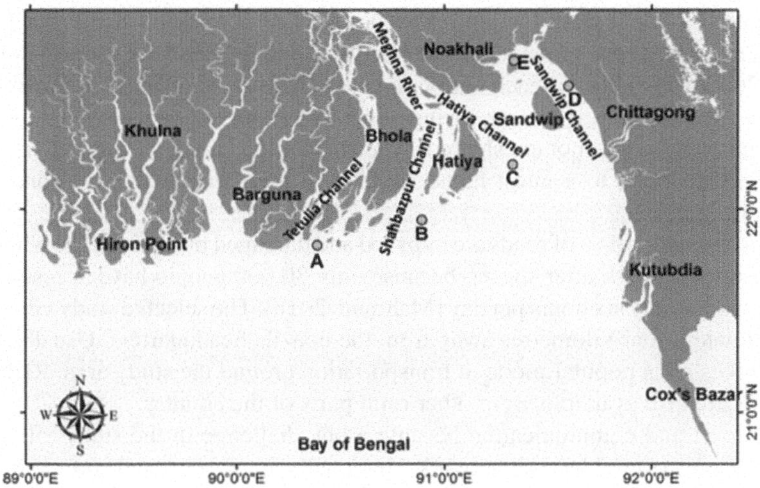

Fig. 3.7 Map of the Meghna Estuary. *Source* Hossain & Momtaz (2024)

Development Board. Later, a mangrove afforestation program was undertaken by the Forest Department to strengthen the embankment, as well as to protect the island in the 1960s. Community-based cyclone shelters, mainly for emergency evacuation purposes, have been constructed since the 1960s. According to the upazila administration, up to the present, the government and other organisations have built 132 cyclone shelters on this island.

Many families, with greater economic resources and better educational backgrounds, have migrated from Hatiya Island. The local administration and KIIs reveal that approximately 50% of residents have migrated from the area. Many people have moved to Noakhali, Chattogram, Chattogram Hill Tracts, Dhaka and other areas. People have left to seek better economic and professional opportunities. Moreover, thousands of families have left Hatiya because of various disasters related to climate change, and these people have often become helpless and vulnerable. There are many uncertainties related to climate change and the magnitude of rising sea levels. These problems are likely to continue and worsen in the future.

3.4.5 Researcher's Observation of the Study Area

This section provides an overview of the study area, based on observations by the researcher and the logistics support needed to reach the study area. As mentioned previously, the study area is located in the south-central part of the country and is exposed to the sea. In addition, this subdistrict is remotely located, and it is one of the most disadvantaged areas in Bangladesh. To travel to the study sites, various modes of transportation were used (auto rickshaw, ship, motor bike and walking). From the capital of Bangladesh, an auto rickshaw was used to reach Sadarghat Launch Terminal, from where ships depart for Hatiya Island. It then took around 9–10 h to travel 260 km by ship to reach Hatiya Island. There were alternative travel means using buses and boats. However, owing to the uncertainty of boat services and travel times, this option was not considered. After reaching Hatiya, a local auto rickshaw was hired to travel to a guest house, used as a base from which to conduct the fieldwork.

Approximately 90% of roads are unpaved and damaged on Hatiya. The whole area usually remains dark after sunset, because only 30,000 people have access to electricity, and for only six hours per day (Mahmud, 2018). The selected study villages are approximately four kilometres away from the upazila headquarters. Usually, rented motor bikes are a popular mode of transportation around the study area. Rickshaws and vans are also available, as in other rural parts of the country.

Transport and communication became a big challenge in the study villages, as they were not covered by road networks. Inter-village transport and communications were mainly conducted along mud roads, and it was therefore very difficult to move about during the rainy season. This issue was mentioned by the KIs and also came up in focus group discussions (FGDs) with participants. People suffered much because of various adverse situations in the area. Soil characteristics and salinity were both responsible for turning soil into mud during the rainy season. Respondents reported that all existing roads and embankments were damaged by the effects of previous cyclones and storm surges. After devastating events, the local administration, with the support of NGOs and local residents, tried to renovate some roads and parts of the embankment. Such efforts were generally not adequate to restore the many affected areas.

After conducting fieldwork in the selected two villages, it became apparent to the researcher that these villages had been frequently and severely affected by climate change-induced disasters in the past. The local residents were under pressure from multiple climate change-related disasters and these might be triggered in the near future. According to the KIs, islands such as Hatiya are vulnerable to future severe river or coastal erosion, rising sea levels, cyclones and storm surges. That is why preparedness is required to tackle the situation; otherwise, the entire island will disappear into the Bay of Bengal in the future.

3.5 Data Collection Techniques

For this study, fieldwork in Bangladesh was completed in two phases. The first phase was conducted from 6 November 2018 to 28 January 2019. The second phase was completed from 15 September to 15 October 2019.

3.5.1 Document Review (Secondary Data and Information Collection)

The relevant literature was reviewed mainly from online sources, as well as published documents. Secondary sources of information came from published and unpublished research related to the study topic. These included books, journal articles, research reports, government reports, NGO documents, conference proceedings, locally published reports, maps and official websites. Moreover, study materials were searched from libraries, databases and relevant literature. At the same time, the researcher contacted various GOs and NGOs to collect relevant reports and publications. This literature focused on climate change governance issues, the roles of GOs and NGOs in climate change governance, the identification of criteria for evaluating climate change governance, community participation in climate change governance, and effects on local livelihoods. To complete the literature review, associated documents on policies, plans and programs were collected, studied, assessed and summarised.

3.5.2 Primary Data and Information Collection

3.5.2.1 Semi-structured Questionnaire for the Heads of Households

The primary data and information were collected from selected heads of households using semi-structured open-ended questionnaires. The heads of households were selected based on gender, age group, and livelihood.

These households were vulnerable to climate change effects in many ways; moreover, they are living in those areas where climate change-related projects have been implemented or are ongoing. The respondents were selected based on a random sampling method. Face-to-face personal interviews provided the opportunity to gain in-depth information from respondents. These interviews provided information regarding climate change governance and its effects on local people's lives and livelihoods. Because of the respondents' level of education, poor communication networks and economic issues, the questionnaire surveys were not administered by telephone or via social media.

The researcher first read an information statement to the heads of households. The participants were then given a consent form to sign. The heads of households who gave their consent to participate then proceeded to interviews. The types of interview questions (open- and close-ended) were developed targeting the background of the respondents and their roles and perceptions of climate change governance issues. Community perceptions of climate change governance were calculated using a Likert scale (Likert, 1932), a 5-point scale ranging from 'strongly agree' to 'strongly disagree', as demonstrated by numerical values of 1 ('strongly agree'), 2 ('agree'),

3 ('neutral'), 4 ('disagree') and 5 ('strongly disagree'). This scale is highly effective in behavioural research deployed for qualitative components.

3.5.2.2 Pre-testing the Questionnaire

The household interviews required approximately 20–30 min to complete. At the beginning, pilot tests of the interview were arranged with a few respondents (e.g. 5 householders) to justify the study's relevance to Hatiya Upazila. This helped to avoid confusion, ambiguity, duplication and inadequacy before the actual survey. The questionnaire was modified based on the feedback from the pilot tests. This approach ensured that the views of householders regarding their participation in climate change governance practices, in the context of their socioeconomic and environmental conditions, were accurately captured. Information provided by the heads of households was then used to select participants for the FGDs and identify KIs for the in-depth interviews.

3.5.2.3 Sample Size Determination for Household Survey

Sampling involves the selection of a subset of individuals from a larger group or population. The two main benefits of adopting sampling techniques in research are faster data collection and cost reduction (Cochran, 1977). The sampling process aims to select a proper population and then gain their participation to obtain accurate field-level data (Blair, 2017). The sample of heads of households was randomly selected from the population of the selected villages of Hatiya Upazila in the Noakhali district of Bangladesh. The sample size was calculated using the following formula (Yamane, 1967), based on the number of households in the study villages, with a precision level of seven percent.

$$n = N/\left(1 + N \times e^2\right)$$

In the case of this study, n = sample size (number of households chosen to interview); N = total households in the study areas; e = level of error.

To minimise the risk of presenting a required sample size determination, a 95% confidence level and a precision level of seven percent were used. Based on the population of the selected villages, the proposed sample size was 181. According to Kish (1965), 30–200 respondents are sufficient to obtain the required number of responses for a field survey. Here, the number of participants recruited in this research meets that objective. This study used a multi-stage random sampling technique. Research was conducted based on invitation and voluntary consent in the selected study villages. This maintained an equal chance of participation from each respondent to join the survey. The householders' data and information supported to select the major livelihood groups and the participants having enough knowledge to be interviewed in FGDs sessions, and assisted to choice key informants for in-depth interviews.

3.5.3 *Key Informant Interviews*

The KIIs were conducted through face-to-face interviews with officials from GOs and NGOs using a semi-structured questionnaire. The KIIs were arranged to obtain in-depth views, experiences and recommendations about the present climate change governance practices of both GOs and NGOs. This method supported the collection of in-depth information from selected KIs, and filled the data and information gaps that existed following the head-of-household interviews. To conduct any research, the minimum number of key informants suggested is 12 (Onwuegbuzie & Collins, 2007). Individuals with adequate knowledge of climate change governance issues were recruited for the interviews. Based on the key issues, a total of 34 key informants were interviewed from the national to the local level (see Table 3.4). Among the KIs, 27 informants were selected from the national level and the rest (7) were identified from the local level.

The KIs at the national level included government officials, representatives of NGOs implementing climate projects, experts from public universities, representatives of research organisations and consultants directly linked with policy processes. At the local level, the KIs include people such as local government officials, representatives of NGOs implementing climate projects, political representatives, journalists, social workers and local knowledgeable persons.

All interviewees were required to sign a consent form to ensure their formal agreement with the process. The interview sessions were arranged in suitable locations in consultation with the participants, according to their availability and suitability. The duration of each interview was approximately one hour, during which respondents also revealed their roles in their organisations in relation to climate change policy, planning, management and implementation. Interviews were conducted in English or Bengali, as per the interviewees' preference. Each KII was assigned an ID number

Table 3.4 Key informants

No.	Key informants for interview	Numbers
1	Government officials	9
2	Representatives of NGO	8
3	Experts from public universities	4
4	Experts from research organisations	2
5	Experts on climate change issues	5
6	Local knowledgeable persons	6
	Total	34

for anonymity in the text of this book. The views of KIs are cited in the chapters using their ID numbers, such as 'KII20', which is the ID number of KI 20. The list of key informants is attached in the appendices.

3.5.4 Focus Group Discussions

The FGD is a process through which a group of participants work collectively on a particular issue to express joint ideas and recommendations related to that time and place (Barbour & Schostak, 2005). FGDs, therefore, help create consensus. As part of this study, FGDs were arranged to gain in-depth information from a selected group of participants.

For this study, FGDs were organised after the reconnaissance survey, household survey and KIIs, to collect and review the information arising from these. People were invited to participate in the FGD sessions and were also requested to sign a consent form.

FGD participants were required to have adequate knowledge to understand and discuss the research issue. People from various professional groups, such as GO officials, NGO officials, local representatives, community leaders, teachers, social workers and community people were invited to participate in the FGD sessions. Each group included six to eight people. At the beginning of a discussion, an overview of the research was presented to participants. The questions used to direct FGDs were similar to those in the questionnaires for all FGD sessions. An open-ended FGD checklist was presented: some questions from KIIs were repeated to maintain consistency, validity and reliability of the findings.

The FGDs covered various issues and provided an opportunity to introduce a number of research tools. The FGDs incorporated the following elements:

- an introduction to the researcher, research assistant, the research project and the participants taking part
- a discussion regarding ethics: informed consent and voluntary joining, confidentiality, the use of data and information, and scope to make enquiries
- open-ended questions and group discussions.

Figure 3.8 presents the entire research process undertaken in this study. In-depth KIIs and FGDs were the major sources of qualitative data. A few interviews and discussions were conducted in Bengali, as this is the dominant language of the people of Bangladesh. The conversations were recorded by digital audio recorder. Note-taking was also conducted to provide a backup source of responses. The results obtained from different sources were cross-checked to maintain validity and ensure the quality of data and information.

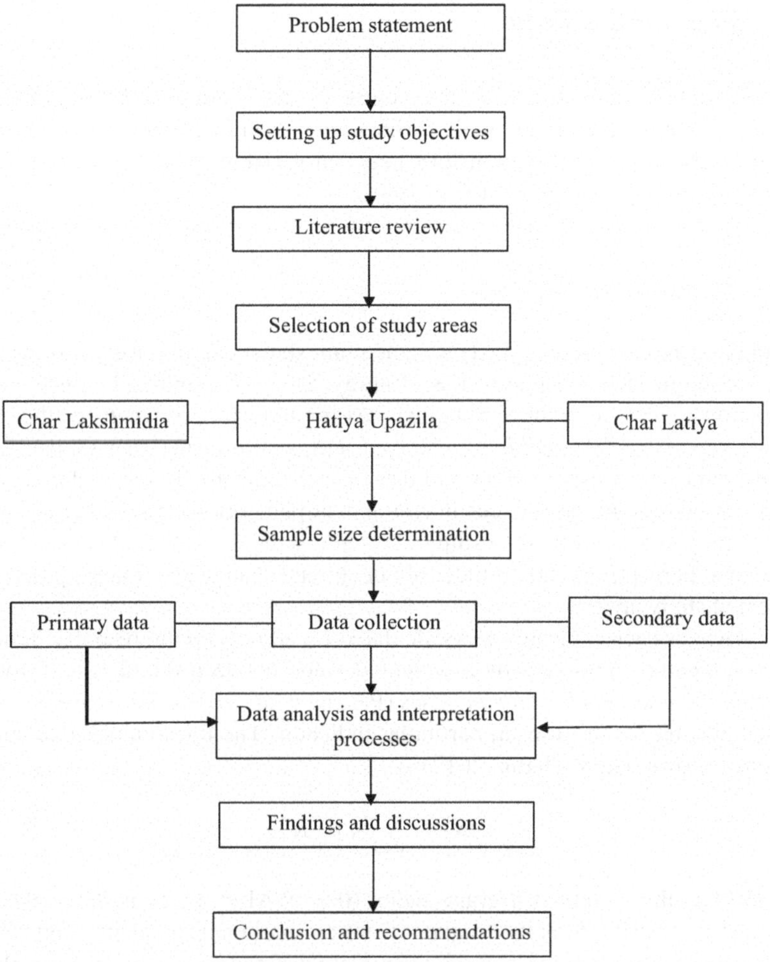

Fig. 3.8 Research design flow chart

3.6 Data Analysis

Data analysis transforms data and information into findings (Patton, 2002). As stated above, this study applied MMR, and incorporated both quantitative and qualitative data analysis methods. The subsequent actions were undertaken during the data analysis involving both quantitative and qualitative techniques. Basic descriptive statistics, combined with qualitative information collected through qualitative techniques (e.g. FGDs) were used to describe and interpret patterns in the data.

3.6.1 Quantitative Data

Quantitative data analysis varies depending on the nature and extent of variables. For example, univariate analysis illustrates prominent features of variables, bivariate analyses present relationships between variables, and multivariate shows relationships among variables (Sufian, 1998).

3.6.1.1 Weighted Average Index

The weighted average index (WAI) was the main statistical tool used, following the scaling technique (Khongsatjaviwat & Routray, 2015). It combined choice weights and question weights to produce a single index for all responses, analysed the checklist scores and computed a weighted average for a questionnaire (Pal & Ghosh, 2017). The WAI identified the degree of importance of each indicator; it is considered a quick technique to assess the differences in respondent perceptions (Pakzad et al., 2016). It was employed to convert the ordinal data into scale data, to analyse the level of respondents' perceptions and attitudes about climate change governance practices in the selected study area.

The WAI for each indicator was calculated by adding up the response numbers for each indicator, multiplied by a weighted value between 0 and 1, and dividing the sum by the number of total responses (Pakzad et al., 2016). This gave an overall weighted average score for each particular indicator. The formula of calculation of WAI is as follows (Ha & Thang, 2017):

$$\mathrm{WAI} = \sum \mathrm{S_i F_i / N}$$

where WAI is the weighted average index ($0 \leq \mathrm{WAI} \leq 1$), S_i is the scale value assigned at its priority, F_i is the frequency of household's respondents, and N is the total number of observations. These indices were designed based on social scale; the value of each index was kept from 0 to 1. The type of each index is described as follows.

3.6.1.2 Perception Index

A perception index was applied to evaluate the level of heads of household perceptions of climate change governance practices in the study sites. It includes five levels (see Table 3.5).

Table 3.5 Perception index levels

Categories	Very good	Good	Medium	Poor	Very poor
Scale	1	0.75	0.5	0.25	0

3.6.1.3 Satisfaction Index

The formula for calculating satisfaction index is as follows:

$$\mathrm{WAI} = (1.00 \cdot f_1 + 0.75 \cdot f_2 + 0.50 \cdot f_3 + 0.25 \cdot f_4 + 0 \cdot f_5)/N$$

WAI is the weighted average index ($0 \leq \mathrm{WAI} \leq 1$); f_1 is the frequency of first scale choice; f_2 is the frequency of second scale choice; f_3 is the frequency of third scale choice; f_4 is the frequency of fourth scale choice; f_5 is the frequency of fifth scale choice. The overall assessment (OA) is calculated from the average of WAI value (Nooriafshar et al., 2004).

3.6.1.4 Quantitative Data Analysis

IBM SPSS Statistics 24 software was used to analyse the quantitative data collected through household surveys to evaluate climate change governance practices at the community level. The raw data were analysed using a spreadsheet (Excel) to summarise the results. Moreover, simple statistical techniques were performed using Excel to demonstrate the data in graphic form; this helped to interpret the results based on the response. In analysing quantitative data, descriptive statistical analysis techniques were incorporated for further quality development.

3.6.2 Qualitative Data

The recorded interviews (from KIIs and FGDs) were transcribed into textual form in the original language of Bangla, along with other data and information. The transcribed files were then translated from Bangla to English. In analysing qualitative data, content and contextual analysis including text, procedural and thematic aspects were used.

3.6.2.1 Application of NVivo

NVivo12 software facilitates the management and analysis of qualitative data (Bryman, 2012). It is a popular software for analysing data in qualitative research (Kabir, 2012). It manages data and ideas, queries data, visualises data, and provide reports from data (Bazeley, 2007). Wickham and Woods (2005) note that NVivo software supports the presentation of research and analysis processes in a more transparent and logical manner. Through this software, nodes and strings were developed depending on the narrative text, such as interviews and notes from the field study. Various systematic analysis and logical interpretation of data through analysis produce a better understating of the themes and issues in the relevant research.

3.6.2.2 Data Process and Analysis by NVivo

Initially, all transcribed files were uploaded in the form of text files to the NVivo12 program. Coding was undertaken, a major task in analysing data in NVivo. After coding was completed, it was straightforward to organise and understand the outcomes. The materials were analysed based on the research questions, to obtain an overview of the results. The processes of interview data analysis using NVivo12 software are illustrated (see Fig. 3.9).

Interviews and discussions were then referenced in this book using an identifying number, such as KII20 or FGD4, to present the respective respondents' comments.

After the analysis of qualitative and quantitative data following various methods, the results were compared and contrasted for different aspects of the variables. This is presented further on by critical discussion and interpretation.

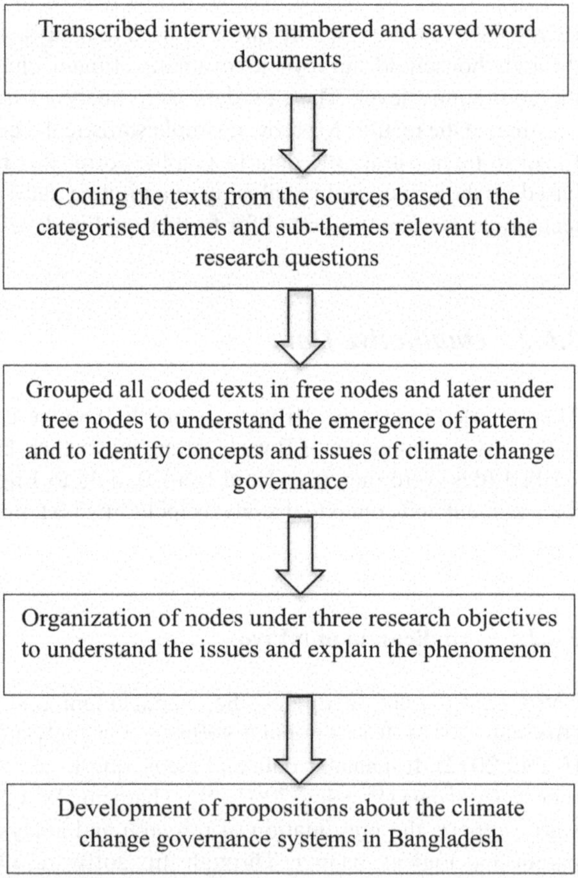

Fig. 3.9 Interview data analysis process using NVivo12. *Source* Modified from Kabir (2012)

3.7 Reliability and Validity

The terms 'reliability' and 'validity' appear almost synonymous, but they differ in meaning regarding the evaluation of concept measures (Bryman, 2012). Both validity and reliability are essential to maintain the accuracy of research. Various techniques such as triangulation, feedback reporting, peer examination etc. were conducted to maintain validity and reliability of the research (Davison, 2014; Noble & Smith, 2015).

3.7.1 Validity

Validity maintains the trustworthiness, utility and dependability that the researcher and various stakeholders are involved in together (Zohrabi, 2013). Validity is necessary to ensure that any particular research investigates what the researcher pretends to investigate (Nunan, 1999). Validity is an important criterion for evaluating the level of quality and acceptability of research (Burns, 1999). It is important that the data and instruments used in research are validated. Creswell and Miller (2000) note that validity is often affected by a researcher's views of validity in the research, and depends on the choice of paradigm assumptions. Golafshani (2003) defined validity as a kind of qualifying check or measure for research.

When research questions are valid for the expected outcomes and the methodology is suitable for answering the research questions, then the research design is appropriate for the methodology. The sampling and data analysis is conducted accordingly and finally the results and conclusions are valid for the expected results (Leung, 2015). Researchers have to use multiple strategies to review the results' accuracy (Cameron, 2009; Creswell, 2013). Various procedures can be adopted to validate instruments and data, such as content validity, internal validity (triangulation, observation, participation), utility criterion and external validity (Zohrabi, 2013).

The questionnaires used in this research were developed to interview the heads of households, based on the survey questions adopted in previous research, and the climate change governance issues considered in the two study sites. To conduct the FGDs, an appropriate number of participants was selected from the targeted persons. Thus, content validity and external validity were confirmed. With the semi-structured interviews, external and content validity were established by the sample size used, the selection procedure, the overall coverage of the concepts, and the relevant issues.

3.7.2 Reliability

The reliability of data and information is considered a main requirement for any research. Reliability deals with the consistency, dependability and replicability of the results obtained from research (Leung, 2015; Nunan, 1999). As data were collected from different original sources, the researcher verified their accuracy in terms of form and context. Reliability was attained through using several procedures to design questionnaires, select participants, conduct the interviews and focus groups, and data analysis (Bryman, 2012). In this study, the questionnaires were developed considering the research questions and pre-tests were conducted to maintain reliability. For the KIIs, the schedules were fixed first, and then a set of questions was given to respondents. With the FGDs, discussions were arranged by contacting the targeted participants and a facilitator then conducted all sessions, thus maintaining research reliability.

3.8 Ethical Considerations

Ethical issues can be seen as part of 'doing no harm' (Piper & Simons, 2005). This study was approved by the University of Newcastle's Human Research Ethics Committee (HREC). A formal application was lodged (Reference No. H-2018-0361) and several revisions were undertaken before gaining final approval. To help ensure the participation of intended people, ethical considerations were followed to avoid participants perceiving problems and to protect their rights at every step of data collection. Ethical matters such as anonymity and confidentiality of data and information, and respondent details were managed properly according to the consent and ethics approved by the University of Newcastle's HREC.

This study was conducted using questionnaire surveys, KIIs and FGDs, so local people's participation was ensured at all levels of the data collection process. Participants who joined the interview and discussion sessions were informed properly about the benefits of participating in this study. Their formal consent was sought to ensure their participation. Participant anonymity for the interviews was protected. Their names were not recorded on the questionnaire, and thus they are unable to be identified. With the KIIs, participants signed consent forms and their permission was needed before they could be audio-recorded. Respondents had the right to not participate in interviews at any time without any need for justification. Participants in the FGD sessions also signed consent forms, and consent was also given to record the whole discussion session. To maintain confidentiality, the resulting data (both hard and electronic copies) were not disclosed to anyone except the research team.

3.9 Procedures Used

The research design followed several specific procedures. A primary investigation was conducted to select appropriate study sites for the research in Bangladesh. Written consent was received from the local upazila administration for their participation and approval to conduct the fieldwork. Questionnaires were developed simultaneously based on the literature review and local context.

For the fieldwork, international flights were used to travel from Sydney, Australia to Dhaka, Bangladesh via Kuala Lumpur, Malaysia. From Dhaka, a passenger ship was taken to travel to the study site of Hatiya Island. A rented vehicle was used in Hatiya Island to conduct the fieldwork as there was no scheduled public transport system on the island. Local administration helped by providing a motor vehicle during the reconnaissance survey. Prior to departing from Hatiya Island, contact was made with relevant local government and non-government offices.

The local language was not an obstacle for this research while conducting the fieldwork. The fieldwork and data collection were conducted in Bangladesh, using the local language Bangla and English. The researcher is from Bangladesh and is a fluent speaker of Bangla. Therefore, no problems were encountered while communicating with the interviewees. All research documents, including questionnaires, consent forms, information statements, and all fieldwork tools were translated from English into 'Bangla' so that the participants could understand them.

3.10 Conclusion

This chapter described the methods used to review climate change governance practices in Bangladesh. It presented an overview of the study area, including the background of various past climate change-related events. It explained in detail the research methodology used in this study. It illustrated why this research design was undertaken, and the sources of information and required tools to conduct this study. Moreover, this chapter also highlighted the various methodological approaches used, such as sampling, designing questionnaires, data collection and data analysis. This chapter also focused on validity and reliability to maintain the accuracy of research. Further, ethical issues relating to research were also explained to highlight the research's confidentiality measures.

Chapter 4 presents the results and discussion regarding climate change governance practices of GOs in Bangladesh.

References

Alam, M. E. (2014). *A new record and geography of tsunamis, earthquakes and tropical cyclones in Bangladesh and their relevance to disaster risk reduction* [Unpublished Ph.D. thesis], School of Biological, Earth and Environmental Sciences, Faculty of Science, University of New South Wales (UNSW), Australia.

Ali, A. M. M. S. (2014). Union. In I. Islam (Ed.), *Banglapedia: National encyclopedia of Bangladesh*. Asiatic Society of Bangladesh.

Ali, A., Mynett, A. E., & Azam, M. H. (2007). Sediment dynamics in the Meghna Estuary, Bangladesh: A model study. *Journal of Waterway Port Coastal and Ocean Engineering, 133*(4), 255–263. https://doi.org/10.1061/(Asce)0733-950x

Bangladesh Bureau of Statistics. (2015a). *Bangladesh population and housing census 2011: Community report (Zila: Noakhali)*. www.bbs.gov.bd

Bangladesh Bureau of Statistics. (2015b). *Trends, patterns and determinants of marriage in Bangladesh*. http://203.112.218.65:8008/WebTestApplication/userfiles/Image/PopMonographs/Volume-13_MM.pdf

Banglapedia. (2023). Hatiya Upazila. In S. Islam (Ed.), *Banlapedia: National encyclopedia of Bangladesh*. Asiatic Society of Bangladesh.

Barbour, R. S., & Schostak, J. (2005). Interviewing and focus groups. In B. Somekh & C. Lewin (Eds.), *Research methods in social sciences* (pp. 41–48). Sage.

Bazeley, P. (2007). Perspectives: Qualitative computing and NVivo. In P. Bazeley (Ed.), *Qualitative data analysis with NVivo*. Sage.

Blair, A. A. C. (2017). *Climate change threats and adaptation responses in small island developing states: A comparative analysis of Antigua & Barbuda and Vanuatu* [Unpublished doctoral thesis], University of Newcastle, Australia.

Bryman, A. (2012). *Social research methods* (4th ed.). Oxford University Press.

Burns, A. (1999). *Collaborative action research for English language teachers*. Cambridge University Press.

Cameron, R. (2009). A sequential mixed model research design: Design analytical and display issues. *International Journal of Multiple Research Approaches, 3*(2), 140–152.

Chowdhury, M. T. H. (2015). Hatiya Upazila. In I. Siajul (Ed.), *Banglapedia: The national encyclopedia of Bangladesh* (2nd ed.). Asiatic Society of Bangladesh.

Chowdhury, S. Q., & Chowdhury, M. H. (2014). Char. In S. Islam (Ed.), *Banglapedia: The national encyclopedia of Bangladesh*. Asiatic Society of Bangladesh.

Cochran, W. G. (1977). *Sampling techniques* (3rd ed.). Wiley.

Creswell, J. W. (2013). *Research design: Qualitative, quantitative and mixed methods approaches*. Sage.

Creswell, J. W., & Miller, D. L. (2000). Determining validity in qualitative inquiry. *Theory into Practice, 39*(3), 124–130. https://doi.org/10.1207/s15430421tip3903_2

Das, H. H. (2017). *Hatiya: History and culture* [Bengali]. Hatiya Printing Press.

Davison, T. L. (2014). Phenomenological research using a staged multi-design methodology. *International Journal of Business, Humanities and Technology, 4*(2), 1–9.

Dwip Unnayan Report. (2008). *Dwip Unnayan Sangstha yearly report*. https://dusbangladesh.org/publications/

Dwip Unnayan Songstha. (2016, March 23). *Dwip Unnayan Songstha in disaster*. https://www.youtube.com/watch?v=SOyRSyS1zD8&t=15s

Ghosh, M. K., Kumar, L., & Roy, C. (2015). Monitoring the coastline change of Hatiya Island in Bangladesh using remote sensing techniques. *ISPRS Journal of Photogrammetry and Remote Sensing, 101*, 137–144. https://doi.org/10.1016/j.isprsjprs.2014.12.009

Golafshani, N. (2003). Understanding reliability and validity in qualitative research. *The Qualitative Report, 8*(4), 597–607.

Government of Bangladesh. (2015). *Bangladesh population and housing census 2011: Community report (Zila: Noakhali)*. Bangladesh Bureau of Statistics (BBS).

Government of Bangladesh. (2017). *Proceedings of Upazila inception workshop on integrating community based adaptation Noakhali.* https://www.bd.undp.org/.../Inception%20workshop%20Report,%20Hatiya,%20Noakhali%2

Graff, J. C. (2002). Mixed method research. In A. A. Tashakkori & C. Teddlie (Eds.), *Handbook of mixed methods in social and behavioral research* (pp. 46–66). Sage.

Ha, H. D., & Thang, T. N. (2017). Fishery communities' perception of climate change effects on local livelihoods in Tam Giang Lagoon, Vietnam. In G. Shivakoti, T. Thang, N. Dung, D. Hulse, & S. Sharma (Eds.), *Redefining diversity & dynamics of natural resources management in Asia* (Vol. 3, pp. 111–124). Elsevier.

Hossain, M. F., & Momtaz, S. (2024). Climate Change Governance Practices of Non-governmental Organisations in Bangladesh. *The Dhaka University Journal of Earth and Environmental Sciences, 13*(1), 105–119. https://doi.org/10.3329/dujees.v13i1.77580

Islam, M. R. (Ed.). (2004). *Where land meets the sea: A profile of the coastal zone of Bangladesh.* The University Press.

Islam, M. A. (2008, March 15). Climate change and development risk: Local perspective. *Daily Star*. https://www.thedailystar.net/news-detail-27722

Islam, S. (2015a). Hatiya Upazila. In I. Islam & S. Miah (Eds.), *Banglapedia: National encyclopedia of Bangladesh.* Asiatic Society of Bangladesh.

Islam, S. (2015b). Mouza. In I. Islam & S. Miah (Eds.), *Banglapedia: National encyclopedia of Bangladesh.* Asiatic Society of Bangladesh.

Johnson, R. B., Onwuegbuzie, A. J., & Turner, L. A. (2007). Towards a definition of mixed methods research. *Journal of Mixed Methods Research, 1*(2), 112–133.

Kabir, S. M. Z. (2012). *A critical evaluation of the environmental impact assessment system in Bangladesh using a holistic approach* [Unpublished doctoral thesis], University of Newcastle, Australia.

Khongsatjaviwat, D., & Routray, J. K. (2015). Assessment of capacities of TAOs and CBOs for local development in Thailand. In A. K. Dutt, A. G. Noble, F. J. Costa, S. K. Thakur, R. Thakur, & H. S. Sharma (Eds.), *Diversity and dynamics in resources and urban development: Volume 1: Regional resources* (pp. 501–527). Springer.

Kish, L. (1965). *Survey sampling.* Wiley.

Kumar, L., & Ghosh, M. K. (2012). Land cover change detection of Hatiya Island, Bangladesh, using remote sensing techniques. *Journal of Applied Remote Sensing, 6*(1), 063608. https://doi.org/10.1117/1.Jrs.6.063608

Leung, L. (2015). Validity, reliability, and generalizability in qualitative research. *Journal of Family Medical Primary Care, 4*(3), 324–327. https://doi.org/10.4103/2249-4863.161306

Lieberman, E. S. (2005). Nested analysis as a mixed-method strategy for comparative research. *American Political Science Review, 99*(3), 435–452.

Likert, R. (1932). A technique for the measurement of attitudes. *Archives of Psychology, 22*(140), 5–55.

Maghaddam, F. M., Walker, R. B., & Harre, R. (2002). Cultural distance, levels of abstraction, and the advantages of mixed methods. In A. Tashakkori & C. Teddlie (Eds.), *Handbook of mixed methods in behavioral research.* Sage.

Mahmud, I. (2018, December 22). The development of Hatiya is in talk only [Bengali]. https://www.prothomalo.com/bangladesh/article/1571191

Mayoh, J., & Onwuegbuzie, A. J. (2013). Toward a conceptualization of mixed methods phenomenological research. *Journal of Mixed Methods Research, 9*(1), 91–107. https://doi.org/10.1177/1558689813505358

Ministry of Environment, Forest and Climate Change [MoEFCC]. (2005). *National adaptation programme of action.* MoEFCC, Government of Bangladesh. https://www4.unfccc.int/sites/NAPC/Country%20Documents/Parties/ban01.pdf

Morgan, D. L. (2007). Paradigms lost and pragmatism regained: Methodological implications of combining qualitative and quantitative methods. *Journal of Mixed Methods Research, 1*(1), 48–76. https://doi.org/10.1177/2345678906292462

Nishat, A., Hussein, S. G., Matin, M. A., Molla, A. R., & Tellam, I. (2009). Adapting to climate variability and change in Bangladesh. In T. Devisscher, G. O'Brien, P. O'Keefe, & I. Tellam (Eds.), *The adaptation continuum groundwork for the future* (pp. 15–30). ETC Foundation.

Noble, H., & Smith, J. (2015). Issues of validity and reliability in qualitative research. *Evidence Based Nursing, 18*(2), 34–35. https://doi.org/10.1136/eb-2015-102054

Nooriafshar, M., Williams, R., & Maraseni, T. (2004). *The use of virtual reality in education.* Paper presented at the American Society of Business and Behavioral Sciences, 2004 Seventh Annual International Conference, Cairns, Australia.

Nunan, D. (1999). *Research methods in language learning.* Cambridge University Press.

Onwuegbuzie, A. J., & Collins, K. M. T. (2007). A typology of mixed methods sampling designs in social science research. *Qualitative Report, 12*(2), 281–316.

Ostlund, U., Kidd, L., Wengstrom, Y., & Rowa-Dewar, N. (2011). Combining qualitative and quantitative research within mixed method research designs: A methodological review. *International Journal of Nursing Studies, 48*(3), 369–383. https://doi.org/10.1016/j.ijnurstu.2010.10.005

Pakzad, P., Osmond, P., & Corkery, L. (2016, December 17–18). *Developing key sustainability indicators for assessing green infrastructure performance.* Paper presented at the International High-Performance Built Environment Conference—A Sustainable Built Environment Conference, Sydney, Series (SBE16)

Pal, I., & Ghosh, T. (2017). *Natural hazards management in Asia.* Sage.

Parvin, G. A., Takahashi, F., & Shaw, R. (2008). Coastal hazards and community-coping methods in Bangladesh. *Journal of Coastal Conservation, 12*(4), 181–193. https://doi.org/10.1007/s11852-009-0044-0

Patton, M. Q. (2002). *Qualitative research and evaluation methods* (3rd ed.). Sage.

Piper, H., & Simons, H. (2005). Ethical responsibility in social research. In B. Somekh & C. Lewin (Eds.), *Research methods in social sciences* (pp. 56–62). Sage.

Quadir, D. A., & Iqbal, M. A. (2008). *Tropical cyclones: Impacts on coastal livelihoods.* https://portals.iucn.org/library/sites/library/files/documents/2008-067.pdf

Sarker, M. H., Akter, J., Ferdous, M. R., & Noor, F. (2011). *Sediment dispersal processes and management in coping with climate change in the Meghna Estuary, Bangladesh.* Paper presented at the Sediment Problems and Sediment Management in Asian River Basins Workshop, Hyderabad, India, September 2009.

Sufian, A. J. M. (1998). *Methods and techniques of social research.* University Press Limited.

Tashakkori, A., & Teddlie, C. (Eds.). (2003). *The handbook of mixed methods in social & behavioral research.* Sage.

Uddin, M. N. (2019). *Impact of environmental governance on local livelihoods in forest protected areas of Bangladesh: A study of Lawachara National Park* [Unpublished Ph.D.], University of Newcastle Australia.

Union Council. (2009). *Workplan of union disaster preparedness.*

Union Council. (2019). *Union disaster management committee emergency plan* [Bengali].

Upazila Adminstration. (2005). *Upazila Unnayan (development) profile, Hatiya, Noakhali.* Upazila Adminstration.

van der Roest, J.-W., Spaaij, R., & Van Bottenburg, M. (2015). Mixed methods in emerging academic subdisciplines: The case of sport management. *Journal of Mixed Methods Research, 9*(1), 70–90. https://doi.org/10.1177/1558689813508225

Wickham, M., & Woods, M. (2005). Reflecting on the strategic use of CAQDAS to manage and report on the qualitative research process. *Qualitative Report, 10*(4), 687–702.

Yamane, T. (1967). *Statistics: An introductory analysis.* Harper & Row.

Zohrabi, M. (2013). Mixed method research: Instruments, validity, reliability and reporting findings. *Theory and Practice in Language Studies, 3*(2), 254–262. https://doi.org/10.4304/tpls.3.2

Chapter 4
Climate Change Governance Practices of Governmental Organisations in Bangladesh

4.1 Introduction

This chapter demonstrates how government institutions operate various activities focusing on climate change issues. A semi-structured interview schedule provides answers specific to the research questions 'What are the roles of GOs in planning and implementing various climate change-related projects and programs?' and 'How is the climate change governance issue reflected in various GOs' actions?'. This chapter first discusses the roles of various government departments in climate change-related activities. It then discusses briefly various government climate change-related plans and strategies. The chapter then examines the climate change governance practices of GOs, including their institutional arrangements, accountability, transparency, stakeholder participation, the effectiveness of activities, and resource dependency and fund mobilisation strategies. Further, it describes the successes and challenges of GOs in managing climate change-related activities.

4.2 Governmental Organisations in Climate Change-Related Activities

The term 'state institution' refers to particular formal bodies developed by governments and public services (Rahman & Huang, 2019). Climate change is an emerging, and not a separate sector or issue. It must be linked to all sectoral activities to obtain the expected results. Activities relating to climate change are visible at the government level in Bangladesh and are similar to those in many developing countries (KII21). GOs are responsible at the macro scale for climate change-related works (KII21). Ministries in areas such as environment, agriculture, water, disaster management, education, health, planning, finance and others are working on climate change issues from their own perspective.

Md. F. Hossain and S. Momtaz, *Evaluating Climate Change Governance in Bangladesh*,
https://doi.org/10.1007/978-3-031-98486-0_4

4.2.1 *Ministry of Environment, Forest and Climate Change*

The MoEFCC is the focal point for climate change issues in Bangladesh (see Fig. 4.1) and it works as a secretariat of the UNFCC. At first, the MoEFCC conducted projects independently, but the government began to include other ministries, such as planning and finance. Later, the Economic Relations Division (ERD) of the Ministry of Planning was incorporated to deal with external resources, because of its status as a National Designated Authority (NDA). An expert on climate change in Bangladesh describes the limitations of MoEFCC:

> The MoEFCC officials, however, didn't follow up on issues discussed in the various inter-ministerial meetings in 2015. This ministry didn't perform enthusiastically in various climate change-related issues, and subsequently it didn't act accordingly to kick off the expected results. (KII21)

The MoEFCC did not show much eagerness in collaborating with other line ministries,[1] such as the Ministry of Power, Energy and Mineral Resources (MPEMR) and the Ministry of Agriculture (MoA) on climate change issues, even though they also play important roles. The MoEFCC did not take the MoE to board yet, and MoE did not pay attention to this. The MoA, as a whole, did not show much concern from the very beginning even though agriculture is one of the most vulnerable to climate change. Although this ministry faces challenges, such as a shortage of labour, skilled staff and adequate funding, the ministry is mandated to coordinate climate change-related activities at the national level (KII24).

The MoEFCC is often considered less powerful because it works as a regulatory ministry, and focuses predominantly on sustainability and pollution control. As a result, development-focused ministries, such as the ministries of Finance, Planning, Water Resources, Local Government, and Rural Development and Co-operatives are loath to accept guidance from the MoEFCC. If they follow the MoEFCC's directives, they will most likely need to put more money into projects to maintain compliance (KII22).

4.2.2 *Department of Environment*

The main role of the DoE is to assist MoEFCC in implementing environmental rules and regulations, including those related to climate change (Alam & Rabbani, 2007). The department undertakes initiatives on behalf of the government, such as climate change negotiation, organising events before participating in COPs, and providing basic information on climate change to the prime minister (PM) and other ministers (KII03) (KII04). The DoE is assigned to conduct activities focusing on climate change

[1] The ministries having projects and programs with climate relevance in Bangladesh, such as MoEFCC, Ministry of Water Resources, Ministry of Agriculture, Ministry of Disaster Management and Relief, Ministry of Fisheries and Livestock etc. (GoB, 2018; Vij et al., 2018).

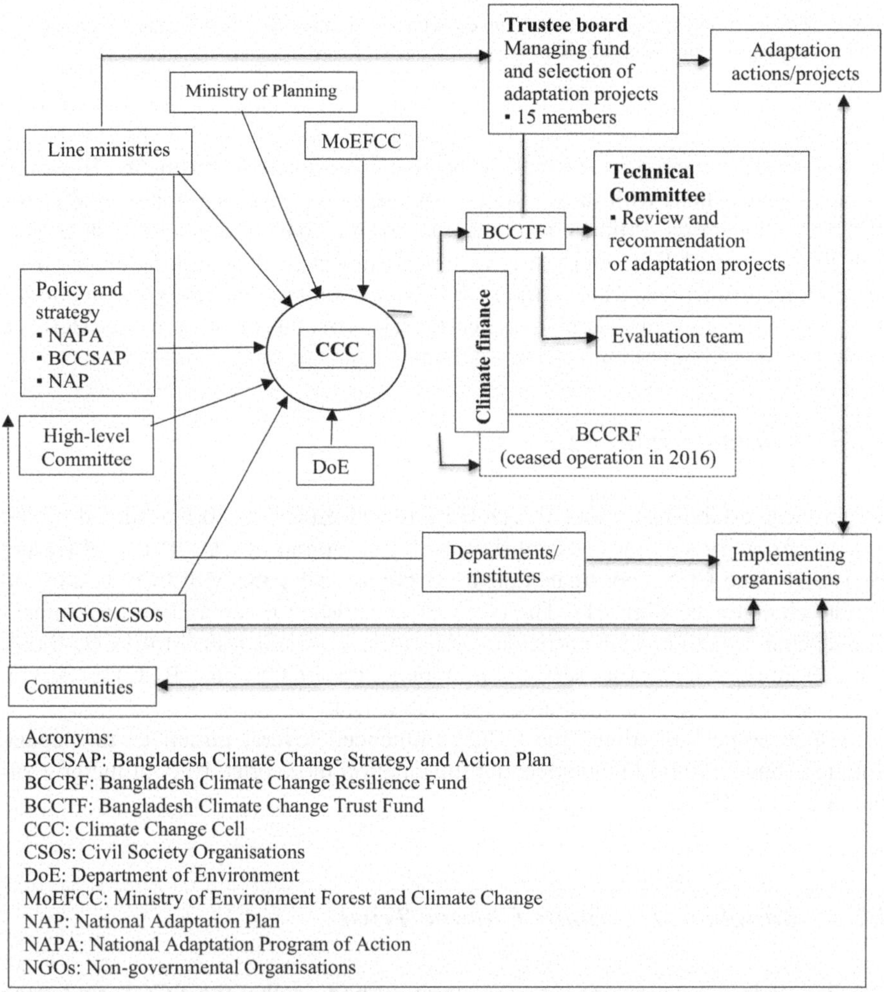

Fig. 4.1 Organisational framework of climate change-related institutions and policies in Bangladesh. *Source* Modified from Ayers et al. (2014), Pervin (2013), and Huq and Rabbani (2011)

effects at the national and international level. Moreover, the department works as a focal point of UNFCC and provides technical support under the ministry (KII01) (KII03) (KII04). The DoE also works to advise other relevant departments on the need to address climate change more effectively (KII04). Moreover, the DoE assesses climate vulnerability at the national level (KII26). Therefore, despite its limitations, the DoE is responsible for conducting activities that focus on climate change. A mid-level official of DoE describes the department's limitations as follows:

> The DoE is facing the shortage of efficient manpower nowadays. This department is smarter in international negotiation but not smart in the field level implementing and monitoring.

So, adaptation and mitigation aren't linked with any individual's issue, rather these are cross-cutting in many aspects. (KII09)

Before starting the CDMP[2] in 2004, training was given to 25 government officials to increase their expertise on climate change issues. These officials would then be recognised as the expert for their respective government departments. However, no meeting was held with these officers, owing to the lack of initiatives after the BCCSAP's inception. Primary training was given as part of a government project with the support of DoE under the MoEFCC; however, this commendable structure no longer functions. That is, the DoE does not want to share its work on climate change with the relevant stakeholders, fearing a loss of control over the issue (KII21).

4.2.3 *Climate Change Cell*

A CCC was established under the DoE by the MoEFCC in 2004 (Ahmed, 2006; Pervin, 2013). This cell was established with the support of CDMP, the DFID and the UNDP. The CCC coordinates all focal points under the MoEFCC relating to climate change (see Fig. 4.1). The cell was established to streamline government climate change activities; in particular, adaptation in Bangladesh (MoEFCC, 2009). The cell delivers secretarial services to manage the BCCRF and BCCTF (Pervin, 2013), and works on four main areas (see Table 4.1).

Despite some limitations, the CCC commenced several initiatives to manage climate change current knowledge and information more effectively at the government level (KII03).

4.2.4 *Bangladesh Climate Change Trust*

The Bangladesh government set up a statutory body called the BCCT, under the MoEFCC, in 2010 (KII01) (KII22) (KII23) (KII27). The body's main purpose is to ensure effective fund use for climate change-related problems. It has also developed clear procedures for reviewing proposals and allocating resources to climate projects. The BCCT is managed by a board of trustees and follows specific procedures to undertake its tasks (see Fig. 4.2). One main task of the BCCT is to implement 44 programs specified under the BCCSAP's six thematic areas. Over the last 10 years, the BCCT has disbursed several hundred million US dollars to a few hundred projects in different government ministries and NGOs (Huq, 2017) (KII09). These projects were implemented by government, semi-government and autonomous agencies; a

[2] The main target of the CDMP was to reduce the vulnerability to natural hazards in the country by integrating disaster risk reduction (DRR) and climate change adaptation (CCA) strategies into the various plans and programs of central, regional and local government agencies (Luxbacher, 2011).

Table 4.1 Major climate change cell activities focusing on climate change

Main areas	Activities
Capacity development	Capacity development of government departments to coordinate and integrate climate change issues in various development activities undertaken by government
Knowledge generation	Better management of climate knowledge and information on impact prediction and adaptation to climate change. Compiling and synthesising of climate knowledge, identifying knowledge gaps and linking information between the science and policy-makers
Awareness raising, advocacy and coordination	Awareness raising, advocacy and coordination with the various partners such as government departments, non-government organisations, civil society organisations, private sector and donor organisations. Integration of climate change adaptation and disaster risk reduction in development initiatives, particularly in climate-vulnerable sectors and disaster risk reduction processes
Improving livelihoods	Enhancing the capacity of livelihoods to adjust to climate change impacts, especially in the agriculture sector. Application of climate modelling in agricultural response options and livelihood adaptation practices

Source Begum et al. (2006)

few were undertaken by NGOs supervised by the Palli Karma-Sahayok Foundation (PKSF). The BCCT began with 82 staff and currently has 62 staff (KII02).

Two major funds have operated in Bangladesh to address climate change issues: BCCTF and BCCRF (see Fig. 4.1). The BCCTF was established by the government from its own sources, whereas the BCCRF was developed from the support of the development partners. The BCCTF is still continuing its functions by disbursing funds to various climate projects of GOs and NGOs, while BCCRF ceased operation in 2016 (Siddique, 2016).

The BCCTF is a GoB self-financing body that addresses climate change-induced disasters (KII01). The Climate Change Trust Act 2010 clearly specifies that funds and their use will be approached in accordance with government rules and business procedures. The Implementation Monitoring and Evaluation Division (IMED) and Comptroller and Auditor General (CAG) offices must ensure that accountability, transparency, and fund use follows these rules, and that climate projects achieve their expected outputs (KII09).

BCCT funds are supposed to be used in a variety of ways. BCCT officials, in association with administration officials, are responsible for checking and due diligence, but debate surrounds what is proper fund use (KII26). While this is a flaw in the laws, as stated in the Climate Change Trust Act, 2010[3]; no revision of the laws has been initiated (KII02). Many projects funded by BCCT were implemented without

[3] According to the Climate Change Trust Act 2010, this trust is assigned to redress the adverse impacts of climate change in Bangladesh by following the rules and procedures.

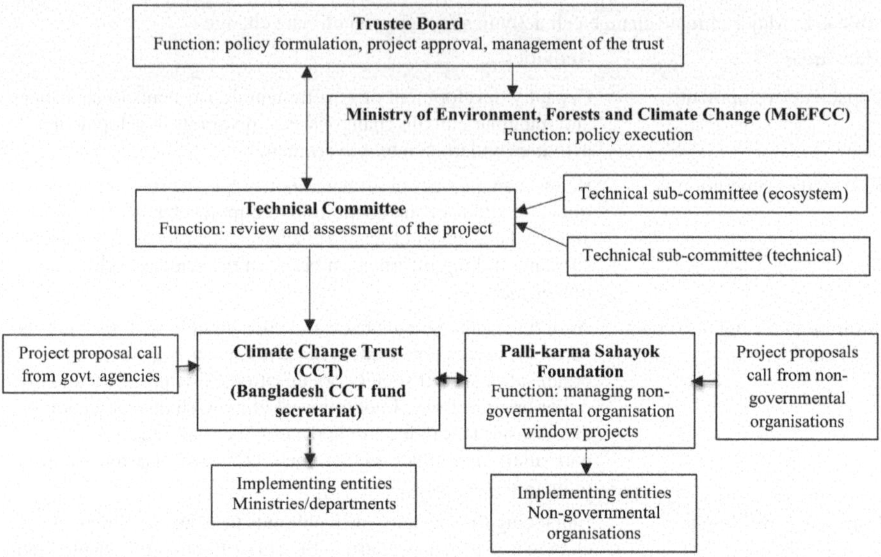

Fig. 4.2 Bangladesh Climate Change Trust fund governance structure. *Source* Rai et al. (2014)

following standard project development processes, and this has raised questions concerning the transparency and accountability of these projects (KII25) (KII26) (Firoz, 2018). The following table summarises the government's budget allocation through BCCT from 2009 to 2010 and from 2018 to 2019 (see Table 4.2).

BCCT had allocated funding for 560 projects up until December 2017 (Ahamed et al., 2018). Of these, 208 projects have already been accomplished (BCCT, 2017), but the evaluation reports of these projects are not universally accessible. These projects were appraised by evaluation committees and approved by the trustee board. The meeting minutes are limited to a few pages; again, these are not open to the public. Therefore, a lack of transparency and public accountability are clearly demonstrated in BCCT-funded projects (KII21).

Unfortunately over time, several instances of fund misuse have been reported by journalists and NGOs such as TIB. Usually, climate funds are meant to address climate change effects, but some instances exist of these funds being used to make environment conservation processes more vulnerable (Uddin, 2018). For example, an embankment was constructed in Barisal district using BCCT funds in 2018. The embankment occupied the river area and narrowed the river flow. TIB (2013) presented various examples where BCCT-funded projects did not follow transparency and accountability issues properly. The following extract illustrates irregularities in BCCT's fund allocation:

> If we see fund allocation of BCCT, there are some irregularities. These funds were supposed to be disbursed in climate change–vulnerable south-western parts of the country, but many projects were granted to Chittogram regions and other parts of the country because of political interventions. Though, these areas were not so much vulnerable for climate change impacts.

Table 4.2 Fund allocation by the Bangladesh Climate Change Trust

Serial No.	Fiscal year	Allocated amount (Bangladesh taka[4]) (in crores[5])
1	2009–2010	700.00
2	2010–2011	700.00
3	2011–2012	700.00
4	2012–2013	400.00
5	2013–2014	200.00
6	2014–2015	200.00
7	2015–2016	200.00
8	2016–2017	100.00
9	2017–2018	100.00
10	2018–2019	300.00
11	2019–2020	152.00
12	2020–2021	67.34
13	2021–2022	99.38
14	2022–2023	24.00
15	2023–2024	83.00

Source Bangladesh Climate Change Trust (2024)

> Various projects were implemented by different government entities, but people don't know the actual scenarios of implementations, and the results. (KII09) (KII27) (KII29)

4.2.4.1 Bangladesh Climate Change Resilience Fund

The BCCRF was an innovative approach between the GoB, development partners (such as the United Kingdom, the European Commission, and Denmark) and the World Bank (WB) to address climate change effects. This fund was managed and used to support the implementation of BCCSAP in 2010 (Finance Division, 2018). The BCCRF was managed by a governing council and a management committee. The governing council incorporated members from government cabinet ministers, donor representatives and civil society, providing strategic guidance to the BCCRF (Finance Division, 2018). The BCCRF disbursed a total of US $71.13 million up until December 2016 (see Table 4.3).

The BCCRF was coordinated primarily by the WB (KII24) (KII27). It allocated funds to projects that were relatively short term compared with the BCCT, and followed strong evaluation protocols; however, its fund distribution systems were quite slow. The BCCRF facilitated millions of people to build resilience against climate change effects (WB, 2012). It did not support any large-scale, long-term projects (Irfanullah, 2020). Moreover, these funds were not mobilised further owing

[4] Abbreviated to BDT.

[5] One crore (BDT) is equal to BDT 10 million.

Table 4.3 Bangladesh Climate Change Resilience Fund investment project disbursement

Project	Total disbursement (US $ million)
Emergency 2007 Cyclone Recovery and Restoration Project	23.06
Secretariat for Bangladesh Climate Change Resilience Fund	0.30
Community Climate Change Project	12.98
Climate Resilient Participatory Afforestation and Reforestation Project	29.89
Rural Electrification and Renewable Energy Development Project II, Solar Irrigation Project	5.00
Total	71.13

Source Bangladesh Climate Change Resilience Fund (BCCRF) Annual Report (2016)

to a lack of proper management, accountability and transparency in decision-making processes regarding fund-handling (WB, 2012).

4.2.5 Implementation, Monitoring and Evaluation Division

The IMED, under the Ministry of Planning, plays an important role in Bangladesh's socioeconomic development through effective implementation monitoring, qualitative evaluation of development projects and ensuring public procurement (IMED, 2008). The IMED strives for result-based monitoring and evaluation, considering the objectives of each particular project. IMED reviews accountability, transparency, fund use and expected returns at the final stage of climate projects (KII21). Sometimes, international third parties monitor projects financed by the Green Climate Fund (GCF) and the Global Environment Facility (GEF) (KII23). Thus, monitoring and evaluation take place in different projects in different ways.

However, prevailing national monitoring protocols and mechanisms are weak in Bangladesh (KII21) (KII22) (KII25). IMED follows traditional procedures to monitor various projects and does not contribute much to successful implementation (KII22). IMED staff follow the guidelines, being mindful of possible discrepancies for various types of projects, when they undergo project monitoring, evaluation and prescription, particularly regarding financial and physical progress (KII23) (KII29).

IMED faces a shortage of adequate staff for their operations because it oversees a large number of projects at the implementation stage (KII06) (KII07). Moreover, IMED staff face frequent transfers after becoming highly competent through training (KII06). The majority of IMED staff are generalists and do not understand the finer points of climate projects, especially technical aspects. Donor-driven projects frequently differ from government projects. Donor-driven projects are often more rigorous and require the involvement of international experts for monitoring and evaluation. These projects are considered more transparent and are well recognised (KII21). However, IMED evaluates these projects and the CAG office also conducts auditing (KII26). With some limitations, the IMED plays an important role for effective implementation of projects, including climate change-based ones.

4.2.6 Department of Disaster Management

The Department of Disaster Management (DDM) was established under the Ministry of Disaster Management and Relief (MoDMR) in 2012 to reduce the loss of life and property. It encompasses various activities, such as fostering sustainable development (i.e. managing hazard-related data), developing disaster management plans and programs, conducting training to enhance stakeholder capacity, and coordinating disaster management activities (Thomalla et al., 2005). Moreover, DDM undertakes activities relating to DRR and disaster risk management. All these activities help manage climate change effects in Bangladesh (KII31). The department faces various challenges, such as a lack of research results to incorporate in policy development, a lack of adequate data, inadequate technological support, and a shortage of resources and skilled staff (KII31).

4.2.7 Non-governmental Organisation Affairs Bureau

The main objective of the NGO Affairs Bureau (NGOAB) is to provide services as part of NGO operations. It also monitors NGO activities to some extent, and ensure accountability to the state (KII13). NGOAB also monitors NGO program implementation at several levels (KII08) (KII09). NGOs must submit progress reports to the NGOAB every year, because issues relating to NGO project finance and outcomes are interlinked. Both NGOAB reporting and donor assessments are key to monitoring capability (KII12). A national NGO managing director describes the NGOAB approval procedure:

> After being selected for a project, we need approval from NGOAB explaining every bit of activities, sources of funds, and allocation of funds in different aspects, and thereby the NGOAB approval is incumbent and elaborate. So, once we obtain approval from NGOAB, then we get funding from INGOs or others. Afterwards, we can proceed for project implementation. (KII09)

NGOAB, having few limitations, coordinates with NGOs, line ministries, government agencies and development partners to reinforce actions relating to climate change.

4.2.8 Palli Karma-Sahayak Foundation

Since 2011, the PKSF has worked with BCCTF and BCCRF under an agreement to mobilise funds to selected NGOs (Rai et al., 2014). PKSF usually receives 10% of the total funds allocated to BCCTF and BCCRF. It is responsible for managing community-level CCA projects through its field-level activities and evaluations (KII08). The following extract explains this:

> We've set up a new CCC to handle the climate funds received from both BCCTF as well as BCCRF. We've implemented Community Climate Change projects from 2012 to 2016 under the GoB with the support of World Bank. We're committed in our activities to support the poor people those who are vulnerable due to climate change impacts at the grassroots. (KII24)

In summary, various government departments are working on climate change issues from their own ground. The organisational framework is already in place and its robustness enables the climate change-related activities more effective. The present organisational set-up is functioning in various ways for the completion of climate change projects and putting impetus to make the vulnerable communities more resilient in climate change-induced disasters. While these organisations have some limitations, they are trying to improve the live and livelihood conditions of the people of this country.

4.3 Plans and Programs on Climate Change Issues

Since the early 2000s, Bangladesh has developed various plans, policies, and strategies focusing on climate change (Stock et al., 2020) (see Table 4.4). Generally, the representation of climate change-related plans and programs in Bangladesh has improved compared with those of many other countries of the world (KII23) (KII30). There are now a number of major policy documents focusing on climate change issues. Sectoral policies have been aligned to include a climate change dimension. For example, the Forest Policy 1994, which is currently being revised, has clearly defined climate change, as has the Agricultural Policy 2010 (draft version). Therefore, Bangladesh has generally made good progress in terms of documentation regarding climate change issues (KII21) (KII25). The main focus of these policies and plans is to improve the country's ability to address and meet climate change-induced disaster issues (KII21) (KII2) (KII23).

4.3.1 Bangladesh Climate Change Strategy and Action Plan

The Bangladesh government formulated the BCCSAP in 2009 following a participation process that incorporated various stakeholders and is currently being revised

Table 4.4 Policies, programs and plans relating to climate change in Bangladesh

Time frame	Policies, programs and plans
1990–1999	National Environment Policy 1992 Forest Policy 1994 Bangladesh Environment Conservation Act 1995 National Environment Management Plan 1995 National Wetland Policy 1998 National Agriculture Policy 1999 National Water Policy 1999
2000–2009	Water Management Plan 2001 Landuse Policy 2001 National Biodiversity Strategy and Action Plan 2004 (updated 2016) Coastal Zone Policy 2005 National Adaptation Plan of Action 2005 Renewable Energy Policy 2008 Agriculture Policy 2009 Bangladesh Climate Change Strategy and Action Plan 2009 Comprehensive Disaster Management Plan I 2004–2009
2010–2023	Bangladesh Climate Change Trust 2010 Climate Change Trust Fund Act 2010 Bangladesh Climate Change Resilient Fund 2010 National Funding Entities 2010 Bangladesh Climate Change and Gender Action Plan 2013 Comprehensive Disaster Management Plan II 2010–2014 National Plan for Disaster Management 2010–2015 Bangladesh National Forestry Policy (Final Draft), 2016 Bangladesh Country Investment Plan for Environment, Forest and Climate Change (2016–2021) Bangladesh Delta Plan 2100 (2014) Eighth Five-Year Plan (2020–2025) National Adaptation Plan of Bangladesh (2023–2050)

(Lopa & Ahmad, 2016). This plan is the first of its type among LDC and developing countries (KII24). The BCCSAP represents the country's commitment to, and understanding of, climate change issues. It focuses on an integrated approach to offset climate change effects (KII22). The BCCSAP is more comprehensive than other plans, and focuses on medium- and long-term actions, such as reducing carbon emissions, adaptation, mitigation, technology development, transfer and financing (Alam et al., 2011). BCCSAP presents a long list of programs under six thematic areas for implementation, to develop individual, community and institutional capacity to cope with climate change's adverse consequences (KII24). BCCSAP is being revised for the next 10 years, taking the country to 2030 (KII30). The following extracts outline some issues relating to the BCCSAP:

> BCCSAP identifies the major sectors only where government needs to focus. BCCSAP is considered as a plan but actually it's not a plan. It describes 76/77 tasks in few pages and no time frame of completion, and all are termed as high priority works but none of them were completed on time. (KII21)

> The updated version of BCCSAP is intended to mainstream climate change-related activities by incorporating all the line ministries and departments for the next 10 years. Now we're going to undertake climate change activities in all sectors comprising national planning, sectoral planning and local planning. (KII30)

The BCCSAP does not describe any quantifiable indicators by which climate change-related initiatives can be assessed (KII24). The plan notes that skilled people are required for proper planning on climate change issues. Eleven ministries were involved in the committee, but they have not considered climate change issues seriously. Instead, the ministries proved they have minimal involvement with climate funds. They placed one-third of funds from BDT 700 crores into fixed deposits in banks, rather than investing in developing skilled staff (KII21).

BCCSAP endorses the climate change responsibilities of stakeholders who are then told to increase their capacity to develop a critical base of expertise in their respective organisations. This has not occurred, and the desired amount of investment has not been implemented at the relevant levels to increase expertise (KII21).

BCCSAP is a country-driven approach and will be implemented through different sectors and their aligned ministries. However, the implementation has not begun. The sole responsibility to implement the BCCSAP rests with the MoEFCC, but this ministry has not performed their task properly. Moreover, the Ministries of Planning and Finance are also responsible, because they are both associated with climate change-related activities (KII08). BCCSAP was the output of a long journey regarding climate change in Bangladesh, and it is considered a well-recognised document. It was developed through the efforts of several stakeholders, mainly government organisations, who not only drafted the policy, but also provided guidelines for its implementation at the field level (Lopa & Ahmad, 2016).

4.3.2 National Adaptation Plan of Action

Bangladesh was one of the first countries among the LDCs to develop a NAPA (Zamudio & Parry, 2016). The NAPA was prepared under the MoEFCC's supervision in 2005. The government delineated the key areas of action on climate change in the NAPA under the UNFCC (MoEFCC, 2005). The NAPA emphasises adaptation measures (Alam et al., 2011). During the NAPA's preparation, six divisional consultations were conducted (KII10). The NAPA focuses on 15 priority CCA activities, but none have been completed successfully. These 15 priority tasks have been allocated funding, but no proper plans were followed, and these are considered scattered works (KII22). NAPA is considered an ad hoc project-based approach in adaptation strategies for Bangladesh (KII30).

The NAPA was updated in 2009 after being criticised by several stakeholders for its limitations in addressing climate change effects. The updated version incorporates various study findings on effects, vulnerabilities and adaptations undertaken previously (MoEFCC, 2009). However, challenges, such as the lack of awareness, tools, knowledge and methodologies were mentioned in relation to the proper implementations of adaptation actions.

4.3.3 *Eighth Five Year Plan 2020–2025*

The Eighth Five Year Plan (2020–2025) contains a full chapter on Sustainable Development: Environment and Climate Change that describes on integrating economic progress, environment, climate change and disaster management concerns into planning and budgeting with the goal of attaining sustainable development. The plan emphasises on investment in green energy, green building, green transportation and green agriculture (Khatun, 2020). The plan focuses governance in relation to achieving targeted goals for climate change issues (KII23). The plan also presents few concerns about climate change that it needs to take on board (KII29). The following table outlines various activities under the Eighth Five Year Plan (see Table 4.5).

Table 4.5 Outlined activities in Eighth Five Year Plan relating to climate change

Issue No.	Activities
1	Improved institutional capacity
2	Activities to mitigate river bank erosion
3	Improved management to control air, water, land and noise pollution
5	Improved coastal afforestation
6	Established forest and carbon inventories
7	Improved co-management in protected areas
8	Improved management of wetlands
9	Improving transnational cooperation on water management
10	Improved data archiving
11	Supporting alternative livelihood
12	Improved watershed management
13	Increased commitment to social forestry
14	Increased commitment to green growth
15	Climate fiscal reforms
17	Mainstreaming climate change and disaster risk management agenda into national planning
18	Accessing international finance for climate action

Source GoB (2020)

4.3.4 National Determined Contributions

The National Determined Contributions (NDCs) determine the national goals for greenhouse gas emission reductions and establish an outline for climate finance mechanisms relating to mitigation and adaptation efforts (Mills-Novoa & Liverman, 2019). In 2015, the Bangladesh government developed NDCs to be conducted by a group of experts under the MoEFCC (with the support of UNDP). This initiative contributed to the Paris Agreement, to restrain global temperature increases to a maximum of 1.5–2°. Despite being a minimal emitter, Bangladesh is making efforts through NDC to reduce their greenhouse gas emissions, under the Paris Agreement. However, a united approach is lacking, and conflicts of interest that exist between different sectors and ministries have resulted in chaotic conditions for NDC implementation. A climate change experts highlights some limitations regarding NDC implementation in Bangladesh:

> There is an absence of institutional set-up to implement the NDCs through required mitigation measures. These are in papers only, but did not take on board yet. (KII08)

4.3.5 Bangladesh Delta Plan 2100

The Bangladesh Delta Plan 2100 (BDP 2100) is a long-term (50–100 years) integrated plan established to address the challenges of water-related issues in Bangladesh. The GoB approved the BDP 2100 on 4 September 2108, to ensure safe living and economic development. The BDP works as an umbrella that integrates all sectoral plans and policies with long-term perspectives. It will enable Bangladesh's policies and institutions to formulate and implement CCA programs more effectively (Choudhury et al., 2012). This plan is a major government initiative that will ensure Bangladesh is safe, climate resilient and prosperous (as far as possible) by 2100. Moreover, the anticipated institutional arrangements, strengthening and capacity building, equity and justice will be ensured in all sectors to achieve good governance (Sarker, 2019). The following extracts describe the BDP 2100:

> This is a good idea to develop this kind of long-term plan to protect this country and its environment from the climate change-induced disasters in Bangladesh. (KII22, KII24)
>
> I think this plan was taken up without conducting much in-depth studies and consultation with the relevant stakeholders. The main challenge lies in its successful implementation. (KII26)
>
> I'm not sure how this country-driven mega plan will benefit us and how this will work, and what would be the expected outcomes in future. (KII27)
>
> Many previous government plans are overlapped with the BDP 2100, and these need to be revised accordingly. (KII22)

However, a number of policy documents, for example, the National Agriculture Policy 2013 under the MoA, contain no specific measures regarding climate change. In the 2018 revision of this document, the drafts were quite effective, but when cabinet approved the policy, they deleted all references to climate change. It is possible for climate change issues to be included in the policy's annex section. This would require a proper action plan (KII24). The following extract explains the plan's present status:

> We're keen to develop various plans and programs to address climate change issues, but very poor in field-level implementation, and that's the sign of poor governance. (KII11) (KII23) (KII25)

These policy strategies are considered rhetorical until enacted. When policies are not translated into action and become integrated with resource allocation, these will not be beneficial at all (KII29). The performances of many organisations are considerably ahead at the planning level for developing plans and policies, but implementing agencies have not yet internalised these approaches, based on the plans and strategies (KII22).

Bangladesh has prepared a number of policy documents on climate change since the late 1990s, such as NAPA, BCCSAP and NDC (see Table 4.4). These plans and programs are considered important tools at the national level to achieve targeted goals on climate change. The achievements of these plans and policies are linked strongly to the proper implementation of climate change programs targeting vulnerable areas and communities. The plans and programs do not often consider climate change effects in different local settings, and these limitations become obvious when the plans are implemented. The plans are considered paper documents only, if the required institutional set-up and coordination mechanisms are not established properly to ensure their implementation. To make these plans and programs more successful, they must be revised based on stakeholder feedback.

4.4 Institutional Frameworks

An institutional setting an appropriate environment to make the climate change-related initiatives more effective. The National Environment Council delivers environmental policies headed by the PM in Bangladesh, and also encompasses members of parliament, government officials and CSO members (see Fig. 4.2). A public university professor describes the council's operations:

> Actually the National Environment Council sits rarely to take decision on environmental issues. This country (Bangladesh) is highly centralised and very authoritarian region in terms of decision taking. If we notice the involvement of PM then it works. Even though it is often difficult to get PM's involvement, the PM is heavily overloaded. Apart from this, beyond the political rhetoric and policy rhetoric, the actual focus in climate change sector is not very strong in this country and it is not government's priority what they say. (KII25)

The GoB has established a National Steering Committee on Climate Change to coordinate climate change-related tasks at the national level. This committee is headed by the MoEFCC and incorporates representatives from relevant government ministries and departments, NGOs and research institutions.

The National Environment Council and the National Steering Committee on Climate Change, in association with the Ministry of Foreign Affairs, also deliver guidance on international climate change negotiations, and provide directions for collaboration, research, exchange of information and development with other countries and organisations. A Climate Change Secretariat has been established in the MoEFCC to support the National Steering Committee on Climate Change. It also coordinates work with climate change units in all line ministries.

See Fig. 4.3 for an organisational framework of climate change-related institutions and policies in Bangladesh. The existing institutional framework on climate change encompasses important issues and shortcomings to undertake necessary actions.

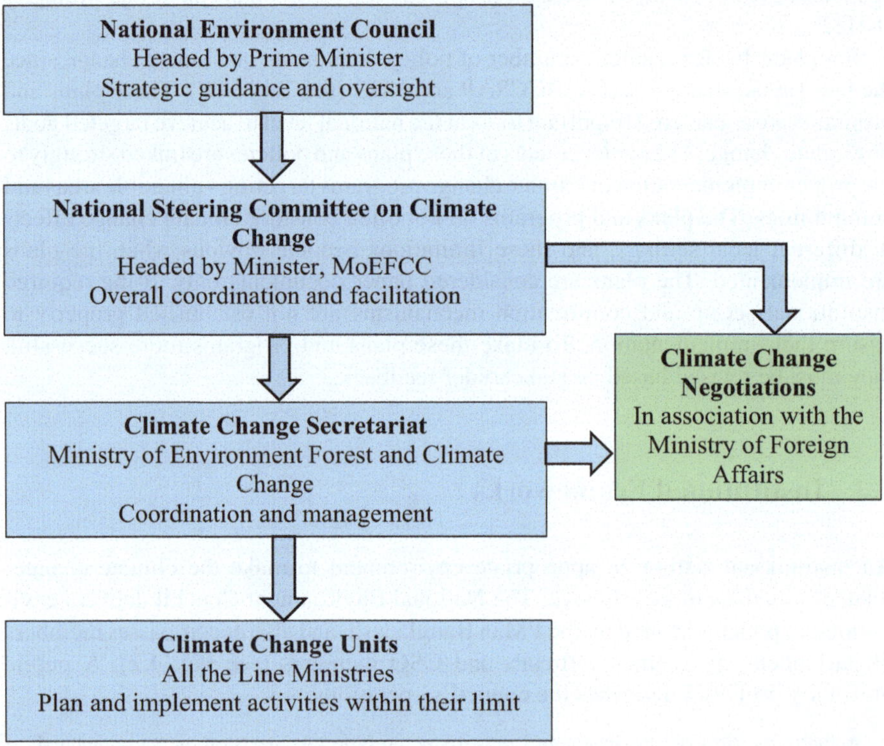

Fig. 4.3 Schematic presentation of institutional structure. *Source* Ministry of Environment, Forest and Climate Change (MoEFCC, 2009)

4.4.1 Strengths of the Institutional Framework

In examining the strengths in the existing institutional frameworks for climate change governance, the following aspects appear constructive for the wellbeing of climate-vulnerable people, despite the many constraints on achieving goals.

An institutional framework for climate governance is just emerging in Bangladesh (KII22) and the government is emphasising the institutional framework issue in its activities (KII23). The institutional framework for climate change incorporates several ministries, departments and organisations in Bangladesh (Rai et al., 2014). The MoEFCC, the focal ministry for climate change issues, is responsible for negotiating with external organisations on climate change issues (KII03). They have overall responsibility for all issues relating to the environment, and act as a focal point through their regulation of the GEF and Adaptation Fund. The ministry has undertaken many measures to increase its capacity to work on climate change-related issues. However, the MoEFCC is not an implementing ministry; rather, it is a regulatory ministry (KII22) (KII29). This situation is explained in the following extracts:

> Institutional framework relating to climate change governance is already took shape in Bangladesh. The government has undertaken various measures for better institutional set-up to deal the climate change issue with additional care. Actually Bangladesh is ahead compared to many other LDCs for setting institutional framework relating to climate change. (KII30)

> We (government organisations) are trying to strengthen our capacity to work on climate change issues. For example, the Planning Commission is also working to establish risk scanning tools for climate change-related programs. Various climate projects are in place and it is an inescapable task to make these projects successful. A well-developed coordination mechanism among the various government ministries and sectors including the MoEFCC, Ministry of Planning, Ministry of Finance, Ministry of Foreign Affairs, MoDMR etc., will promote the ongoing efforts to address climate change issue. An updated BCCSAP, under the process of revision, may outline the procedures for an effective implementation of various climate change-related programs. (KII03)

Government departments such as IMED, under the Ministry of Planning, monitor and evaluate various government projects relating to climate change on a routine basis. They also hire other firms to conduct in-depth monitoring and evaluate the impact of certain projects (KII06). A retired government high-level official explains this:

> Government's Development Project Proposals [DPPs] are now instructed properly focusing on climate change issues. So, on paper, we can see various significant achievements in this aspect (KII26). It is really a good initiative by the government renaming the MoEF as the MoEFCC. This decision indicates that government is putting emphasis on climate change issue. (KII29)

Government departments are now actively participating in climate change-related activities in Bangladesh. Even though the present institutional framework has some limitations, the concerned organisations have adopted measures to make the existing institutional framework more effective. However, some activities relating to climate change at the government level adhere to the present institutional framework. Climate funds are channelled to various government departments through numerous projects and are manifested in the annual budget. Clearly, government organisations are now more connected to address the climate change-related issue than in the past.

4.4.2 *Weaknesses of the Institutional Framework*

The institutional framework for climate change is not currently well-organised in Bangladesh. Some overlap exists in the decision-making processes of several ministries and departments (KII21, KII26). Moreover, institutional inequities and the lack of institutional flexibility are social and practical barriers (Khan, 2019). The following extract explains this situation briefly:

> We've developed several institutions to deal with the climate change issue. However, we often observe various problems among these institutions due to the different landscape and lack of coordination. (KII03) (KII25)

The sole responsibility for implementing the BCCSAP clearly rests with the MoEFCC, but it is not within that ministry's charter to conduct every climate change-related project, nor to formulate every climate change-related policy or protocol. For a successful implementation of BCCSAP, the involvement of other line ministries, such as the Ministry of Planning or the Ministry of Finance is required. Owing to the lack of effective participation from major government departments, the institutional framework relating to climate governance does not function well in Bangladesh. If institutions do not emphasise polices and strategies, then the institutional systems do not function well (KII08) (KII27). Moreover, Bangladesh faces difficulties in the development of its National Adaptation Plan (NAP) (KII08). The following extract explains institutional weakness:

> Institutional weakness is visible in managing climate funds and in monitoring systems. BCCT is using BDT 100 cores (USD 11.8 million) every year and they aren't developing the expertise. In 2013, the Annual Development Program [ADP] was $18 billion, and this was fixed after the negotiation with relevant ministries, but they didn't take the responsibilities and the rate of implementation was really poor. Various measures were undertaken in the initial stage, but the relevant stakeholders weren't so committed in the overall processes. (KII21)

Other serious problems, such as the lack of coordination and institutional competition have arisen, creating further concerns about the controls and protocols for the government's climate change initiatives. When questions arise regarding which ministry or department is responsible for climate change action, this lack of responsibility is observable in the government organisations (KII21, KII25). All government agencies wish to exercise power and some are qualified to do so. Government department performance can become questionable because of the poor management of climate-vulnerable hotspots in Bangladesh (KII03).

Again, conflict exists between the MoEFCC and the Ministry of Planning. Donors are also becoming aligned to different ministries. For example, the German Agency for International Cooperation (GIZ), a German-based cooperation agency, is aligned to MoEFCC, while UNDP is aligned to the Ministry of Planning and the ERD under the Ministry of Finance. The institutional mechanisms are therefore not offering suitable ways of consolidating the approach to climate change. The following extracts explain this institutional weakness:

> The MoEFCC, in the beginning, conducted various projects, but later government included Ministry of Planning and Ministry of Finance to deal the climate change-related issues. Later, ERD, a wing of the Ministry of Finance, was incorporated to deal the external resources as they work as NDA. In the inter-ministerial last five meetings, MoEFCC officials were there, but they didn't follow up the regular discussed issues. Therefore, the MoEFCC didn't want to accompany others to lose the control over the climate change projects, whereas other concerned ministries and departments aren't cooperating in the overall processes. (KII21)
>
> We've too many institutions involved in climate change issue in our country. Now these institutions are facing various problems such as they are different in nature, lack of coordination and very ineffective coordination. (KII25)

The present institutional arrangement will be part of the updated BCCSAP (KII04). Government departments are putting effort into climate change-related projects based on a piecemeal approach right now, but not mainstreaming the climate governance issue in other development planning. Therefore, the existing institutional framework and government mechanisms is not able to produce the expected or desired results in relation to climate change effects (KII08).

4.5 Achievements of Governmental Organisations in Climate Change-Related Activities

Bangladesh has made progress through the effort of government departments and NGOs, progress based on a political commitment to climate change issues. This country, with so many limitations, is often no longer considered a climate-vulnerable country in the world (KII23). Bangladesh has achieved remarkable progress with its CCA measures that are ahead of other LDCs (KII27). A transition is taking place in climate fund mobilisation, as funding opportunities are providing hope (KII23).

The GoB has established a National Steering Committee on Climate Change headed by the MoEFCC (see Fig. 4.2). Other relevant ministries and departments, NGOs and research institutions are also included in this committee (GoB, 2005). Some ministries have developed different capacity levels on climate change issues; many officials have gained knowledge about climate change, enabling them to incorporate new approaches into regular development practices (KII30). The government has incorporated 20 ministries in the Climate Fiscal Framework, and is planning to include 30 ministries by 2021 (KII29).

The CDMP was a good example of this commitment in an earlier period. It was a DRR program, but it addressed CCA issues. Some experimental adaptation programs were taken by the government with the partnership of NGOs, for example solar panels, a floating garden intervention and climate-resilient crops (KII25).

Government departments such as the Planning Commission provide training to officials, enhancing their capacity and incorporating climate change issues in all sectors. This process is continuing. The Planning Commission has incorporated climate change issues into their DPP guidelines and must report on these regularly (KII22).

The Ministry of Agriculture has started to develop an adaptation plan, and the Ministry of Health is working on a similar plan. The MoE is looking for renewable energy sources to reduce dependency on fossil fuels, but non-renewable energy remains unfocused on being environment friendly (KII22). The Ministry of Finance has prepared climate change budget report in each national budget from 2017 to 2018. It allocated nearly 7.5% of the national budget to tackling climate change through 25 ministries across the government in the current fiscal year. The total amount of the climate change budget is almost USD three billion (Huq, 2020).

Under the Rio Project of DoE, government officials are being trained in the Public Administration Training Centre, Regional Public Administration Training Centre, Planning Commission and Planning Academy and across different agencies, so that they can enhance their capacity for developing effective projects. Moreover, they will gain an understanding of the most important aspects or practical details in addressing climate change issues (KII03). In the following extracts, a mid-level government officer and other KIs illustrate this approach:

> As we're implementing various projects, our overall understanding level in climate change issues has improved over a period of time. Our PM is aware of it and PM is very sympathetic to our activities. At the ministry level, having different prioritised issues, they can't emphasise enough the climate change issues. (KII03)
>
> We're working on a citizen climate report. We're working to know the characters of disasters and trying to make people more aware by using the data and information of disasters from 1970 and onwards. We're empowering the citizens by providing various information related to disasters. The citizens are getting aware also and asking for more information. They're interested in getting the budget information and the expenditure. (KII29)
>
> Various government departments are working to acquire and manage data on climate change issues and future scenarios. They are emerging on knowledge based information and getting well-equipped in decision-making processes, so that they can work more efficiently. (KII23)

Bangladesh was one of the first climate-vulnerable countries to develop NAPA. The GoB has allocated BDT 3600 crore (around USD 424 million) from its own resources to implement various plans focusing on CCA and mitigation through BCCT (KII02). The government has spent about one percent of GDP (equivalent to USD two billion) per year since 2010 for adaptation (The Daily Star, 8 Sept. 2020).

4.6 Challenges of Governmental Organisations in Climate Change-Related Activities

The study participants highlighted a number of challenges concerning climate change governance-related issues at the government level (see Table 4.6). Mainstreaming climate change across government departments is a rigorous and slow process (KII25). Moreover, it is a significant challenge for the government to tackle climate change-related problems by making the people resilient and maintain development momentum (KII21).

First, many projects and programs are not yet institutionalised, so their long-term sustainability beyond the project period remains a serious question: these projects are often financed either by the government or donors. As such, beyond the funding period, these results of these interventions are not maintained, despite being interesting and innovative projects (KII09) (KII25).

The second challenge is gaps at the policy, think-tank and grassroot levels in relation to climate change issues. The views of think tanks and researchers, and action plans at the government level do not meet the understanding at a community level (KII24). Moreover, various government organisations, at the local level, are implementing projects to address climate change (KII27). As an example, local vulnerability issues are not comprehensively studied when adopting climate change plans and programs, but local communities expect that local issues will be addressed. Consequently, these persistent knowledge gaps are often responsible for the failure of programs to achieve the expected results. Two senior officials, one from a reputed NGO and another from a research organisation, explain this situation:

> In reality, the mindset of whole government machineries relating to climate change projects aren't inclusive, they don't follow much on climate change-related policies and plans rather they work to maximise their interests. So, whatever the investment for climate change-related projects at least grassroot level people need to get some benefits but these're absent in most cases. (KII26)

> The technical people can predict the research outcomes better, but they can't decide much. On the other hand, policy-makers can't understand the issues properly. The needs of grassroot people are not reflected much in planning and implementation processes. We're acting in a situation of ignorance and couldn't find out the main issues of climate change-related problems. (KII24)

A third challenge is the minimal understanding of climate change issues at the national level. Ministries are not yet competent at including climate change issues in all development sectors. Officials at the ministry level who are involved in climate change-related programs possess a low capacity to deal with the issues. Those with knowledge are not included in climate change implementation at relevant levels (KII21). For example, staff working at the BCCT office who are involved in decision-making processes do not follow multi-criteria selection procedures to approve climate change projects. This raises a question about the capacity of high-level government officials (KII23). Ultimately, they cannot contribute much to climate change programs because of their low-level capabilities, and they are often confused over

Table 4.6 Participant responses to climate change governance challenges

Responses	Sources and examples
Duration of the projects	'The duration of many projects are very short. As the climate variability varies over the years. If we've one year project we can't experience the variability aspect. So we need multi-year projects to see the variability and catch up the climate variability over the years and over the seasons' (KII09) 'We often face difficulties due to short duration of projects' (KII25)
Policy-level gaps	'Government has developed various systems to do the works but which system should be followed is not clear till now' (KII24) 'In case of BCCTF projects, we've gone through the projects, and many projects have no relation with the climate change issue' (KII27) 'However, there are doubts on many climate projects as to whether they are mainly climate change-related or not. So, something is wrong at the planning level' (KII21) (KII26)
Low-level understanding	'Actually the understanding on climate change issues in various government machineries is not well developed yet' (KII03) 'The selection procedures for climate projects are not based on multi-criteria and these raise the question about the associated government high officials' (KII23)
Funds shortage	'We can't allocate funds for all submitted projects, and sometimes we can fund less amount that is not enough for proper implementation of the project' (KII02) (KII03) (KII05) (KII25) 'For climate projects additional funds are required and the government funds are not enough, so we need funds from external sources' (KII25)
Lack of skilled staff	'Majority of the government officials are generalists and they don't have adequate knowledge on the technical issue on climate change' (KII21) (KII25) 'After becoming skilled on a particular issue, we receive posting orders to work on a different issue, so we can't use our expertise' (KII02) (KII03)
Political interventions	'We often observe that climate projects are approved for the less vulnerable areas avoiding the most vulnerable parts. These are happening due to political connections' (KII24) 'The quality of the works deteriorates due to political connections in climate projects' (KII26) (Transparency International Bangladesh, 2013)
Lack of adequate data	'We [government departments] often can't estimate the requirements to work on climate change issues due to lack of required data' (KII22) 'If we ask for a list of climate-vulnerable people in a particular area, the list is not perfect at all. We don't have that type of lists yet' (KII09)
Lack of coordination	'Various ministries don't want to work together for the effective implementation of climate projects' (KII21) 'Repetition of climate change-related activities causes waste of resources and impedes accountability of relevant government organisations' (KII04) (KII24) (KII26)

Source Authors' field survey (2019)

development projects and climate projects (KII03). Many NGOs working on climate change lack experience in this sector and are unable to realise the expected results from their programs. Further, the community does not have enough awareness of climate change issues.

The fourth challenge is inadequate funding for climate change-related projects; this has become a concern related to the sustainability of these projects (KII03) (KII05) (KII25). Bangladesh is also experiencing a general decline in the funding of development projects, with the official announcement of its graduation from LDC to developing country status. Bangladesh will no longer be qualified for the special window of funding it used to receive routinely. The funding of development projects—including climate change projects—has reduced, especially over the last four years. Although GOs claim to have adequate funds for climate change-related actions, they need to generate funds from their own sources (KII05) (KII25). The following extract explains this:

> We can't provide funds to all submitted projects. For example, we have allocated funds for 15 projects out of 80 project proposals. Government allocated BDT 700 crores in the beginning, but in the following years we received BDT 100 crores, and we received BDT 300 crores in the last year. People of different quarters are getting more information about the activities of BCCT at the present time, whereas government is reducing the fund, so this is making a challenging situation for us. (KII02)

The number of projects submitted for funding has increased tremendously over the years (KII02). However, the funds received for climate projects are not enough to ensure these projects' successful completion (KII21).

The fifth challenge is the shortage of skilled staff on climate change issues. Bangladesh's bureaucracy has not yet developed to the point where it can consider climate change properly. Many knowledgeable persons are available to work on climate change issues, but their combined efforts, including those of government officials, are generally inadequate when dealing with various climate projects. Moreover, they are not in a position to take up the challenges involved in climate change issues (KII21).

Many government officials do not have enough technical knowledge on climate change issues (KII21). Their mindset is to only produce development projects. Integrating climate change into a project proposal is a significant challenge. How can mainstream climate change issues be addressed within sectors, sub-sectors and at the ground level? (KII26). The expertise of people working in the BCCT is not at the levels expected for these projects.

A mid-level government officer explains this situation:

> The planning wings in different ministries and agencies aren't competent enough, and it's an inherent problem in the LDCs and developing countries. Now the challenges of the climate change issue have come over. So I'm not blaming these planning wings, but actually they need to improve the quality of staff, and they need some sort of training to deal the climate change issues, so that they can prepare good projects by addressing the issues. We're participating in different international negotiations and forums to work on climate change issues, and we need to improve our capacity. (KII03)

The sixth challenge is an inevitable one face by political hierarchies everywhere: government departments selecting politically affiliated organisations to grant funds to (Bhuiyan, 2015). Dutta-Koehler (as cited in Bhuiyan, 2015) noted 'the relevant government entities chose those projects first for funding which come along with the recommendations of influential ministers'(pp. 358–359).

There are few highly competent people, and they are generally not involved in government systems. Every issue, including climate change, is approached through party politics. Most professionals avoid present political practices because they have their own professional agendas and activities. Most of the promising, efficient and competent people cannot provide national services, and cannot establish themselves in mainstream institutions because of inactive political roles (KII21) (KII23). A senior official of a national NGO demonstrates in the following way:

> Although some government departments try to maintain accountability and transparency in various climate projects, they can't proceed due to political interventions. We don't find any significant move in the party level to take it in a candid way, so that the climate change issue could be addressed with the extra care and interventions. (KII26)

Climate change issues are becoming more politicised and potentially valuable experts lie outside the political arena. In the present governance system, the words of key people are final. A party manifesto is a kind of document declaration. Political parties are committed to implementing their election manifesto. City mayors, *upazila* and union chair people cannot make decisions contrary to political party policies. Additionally, all these organisations are centrally controlled, with bureaucracy dominating every aspect of their operation. The political commitment to climate change is not yet sufficient (KII21).

The seventh challenge is that government departments face a lack of data and proper strategies. Again, climate risk information is not properly updated with current and future climate-related national development policies in Bangladesh (Khan, 2019). For example, cyclone frequency and storm surge heights are increasing in Bangladesh. The country is one of the most well prepared for tackling cyclones, storm surges, floods and other natural disasters. GOs have some capacity to these occurrences, as they follow standing orders on disaster. As an example, embankments in need of repair because of the effects of climate change will need to be constructed at 5–7 m in height. The ministries working on these issues have minimal awareness of the requirements for extra land and funding for these climate-led projects; instead, they focus on standard processes to complete the projects (KII22). This situation is explained by an expert on climate change:

> We're still facing difficulties due to inadequate data on climate-vulnerable areas. People are not getting the right persons and right packages. After the Cyclone Aila, it was reported that the affected people needed to pay for the government support in a few areas. If we have a complete data base of the climate-vulnerable areas, i.e. the number of landless people and their location, we can support them easily and quickly. These datasets need to be updated regularly. If we ask a Union Council office for a list of the vulnerable people, they may provide the partisan lists who usually vote for them. If we can get the accurate list, then we can decide well and proceed quickly to help the victims during the emergency time. (KII09)

The eighth challenge is that climate change is a dynamic issue, and it is difficult to predict with accuracy what will happen in 2030, 2050 or 2100. In-depth analysis is required at the local level regarding what climate change effects will be felt in specific areas, and what the likely scenarios will be will be, particularly in 'hot spot' or agro-ecological areas. Climate risk factors are not considered properly when integrating them into planning and programming processes (KII23).

The ninth challenge is the lack of coordination among departments on climate change issues. Climate change issues are mostly linked with water in Bangladesh. Several ministries working on climate change are not interested in working together to solve problems effectively. Current project formulation practices are not well developed. For example, GOs build roads through waterbodies without including appropriate culverts, but excess water must be drained during the monsoon season. Therefore, inadequate runoff is causing drainage congestion in the monsoon season when another 10 or 15 cm of rain water is added (KII21).

Moreover, different organisations work on the same type of projects. This results in a waste of money and a lack of accountability in the relevant organisations (KII24) (KII26). A high-level official from a NGO explains this challenge:

> If we apply for funds focusing on climate change, we often see that the similar types of projects are undertaken by other departments already. So, the repetition of the projects obstructs in the maximum utilisation of resources and waste of time. (KII04)

There are other challenges for GOs relating to climate change governance practices in Bangladesh. One major challenge relates to Bangladesh progressing to a middle-income from a lower-middle-income country, while simultaneously becoming climate resilient (KII30). The impacts of climate change will result in significant problems for the lives and livelihoods of many Bangladeshi people. It is therefore a major challenge to tackle climate change effects and continue development momentum simultaneously. Moreover, many projects are not succeeding because of poor design and implementation (KII24). There is a shortage of tertiary-level education on climate change issues in Bangladesh. Although the DRR is focused on various academic programs, climate change is not considered in great depth (KII21).

4.7 Governmental Organisation Accountability in Climate Change Activities

Accountability is an important criterion to assess the ultimate performance of governance mechanisms for responding to climate change (Biermann et al., 2010). Although government departments try to maintain accountability in their climate change-related activities, they are unable to proceed effectively because of political intervention (KII26).

In Bangladesh, the National Parliament usually delivers the budget, but the parliament asks for an annual report. The CAG office provides a regular report to parliament, making them accountable to legislators. If climate change receives more attention at the national level, then accountability will be more transparent (KII29). The following extracts explain this:

> BCCT was created in a place of complete mistrust. Actually we couldn't ensure that kind of accountability to make it more functional. For example, BCCT receives BDT 100 crores[6] annually for its operation, and that amount isn't adequate, and can't expect to accomplish every aspect of accountability. Climate change and related accountability aren't integrated into systems yet. (KII21) (KII22)

> But there are many concerns that a large portion of the funds were disbursed from the political context. There are lots of media reports on corruption and misuses of BCCTF, and greater part of the funds were disbursed to the government departments such DoE, Department of Agriculture (DoA), Forest Department [FD] etc. So, who will ensure the monitoring of these projects? (KII08)

Bangladesh's coastal districts (the south-western parts of the country) are the most climate-vulnerable areas. However, around half of BCCT-funded projects were allocated to the Chattogram region, because the former environment minister came from that region. Many organisations have received funds through claiming political identities, and many have no activities at the field level (KII24).

A mid-level NGO staff describes political interventions:

> Our fund receiving possibility is often obstructed by the political interventions. We submit proposals which are worthy to get climate funds from government departments. We can see the political influences in project selection processes. So, we often fail to receive the funds, and this is really unexpected to us. (KII10)

Government departments are failing to conduct planned tasks and are allowing corruption when they should maintain accountability (KII21). For example, the Climate Public Expenditure and Institutional Review 2012 (CPEIR) report, prepared by the Ministry of Planning, emphasised accountability in government-funded climate projects. Government departments neither maintain nor follow any particular accountability protocols or procedures. Those who receive funds from BCCT have some level of accountability, but the requirements are not stringent. BCCT developed some monitoring indicators, such as logical frameworks, project fund limits, project duration, relevance to climate change and environmental impact assessments to endorse during monitoring processes. However, the monitoring of climate change-related issues is not yet fully developed (KII24). The following extract explains BCCT's roles in maintaining accountability:

> We can see the accountability of BCCT like other government organisations in Bangladesh. Our projects are audited by the CAG office. Our projects are monitored by government staff at different levels. Bangladesh Institute of Development Studies [BIDS][7] is evaluating 27

[6] BDT 100 crores is equal to USD 11.8 million.

[7] Bangladesh Institute of Development Studies (BIDS) is an autonomous multi-disciplinary organisation in Bangladesh.

> projects approved by BCCT. We also send letters to the District Commissioners[8] (DCs) to review the progress of BCCT projects, and they also monitor the projects regularly. (KII02)

Regarding BCCT projects, technical committee members accompany BCCT officials to visit project sites. Sometimes field-level staffs deliver videos, pictures and updates of projects. If projects are large enough, then officials from BCCT and relevant ministries will visit to monitor the project's progress. However, in the case of smaller projects (with a budget of less than BDT one crore), less monitoring is conducted. The BCCT improved the monitoring system recently and tare trying to standardise their systems (KII03). The following extract explains the limitations of BCCT monitoring:

> We couldn't succeed to impart a result-based monitoring system like other countries. If you asked about climate change negotiation processes, LDCs didn't agree to put the climate change negotiation under review. We've some weaknesses and we're hoping to overcome these gradually. In that perspective, what BCCT is doing is considered adequate, but these need to improve more based on international standards. (KII03)

Moreover, these projects are implemented mostly by GOs. An important question remains: who will review these projects? Additionally, these projects are not part of traditional Development Project Proposal (DPP) processes. Projects from any government department must go through a planning cell to the Planning Commission; these are then forwarded to the Executive Committee of the National Economic Council (ECNEC) for approval. Contrary to this procedure, BCCT-funded projects are approved by the BCCT's trustee board, so IMED does not monitor these projects; rather, these are monitored by BCCT officials.

BCCT has only a few officials to monitor their more than 100 ongoing projects. BCCT-financed projects are implemented by government officials such as the Upazila Nirbahi Officers (UNO), District Commissioners (DCs) or even secretaries. Project staff, as ordinary government staff, faces difficulties when they attempt to monitor BCCT projects. BCCT-funded projects do not follow normal efficiency and transparency procedures, and are not very effective. Few projects on afforestation programs exist in coastal areas, and the exact condition of those projects is unknown (KII26). The following extract explains the weakness of BCCT accountability:

> Although government staff are trying to maintain accountability in climate projects, but they couldn't follow the implementation processes properly due to political interventions. It has been reported in the newspaper that in the last board meeting of BCCT, around 100 projects were presented for the approval. How will the government staff manage this high number of projects? In the last three years, GCF approved only 76 projects. If we can't maintain these types of mechanisms, these projects will not be more effective. (KII26)

PKSF is considered one of the most efficient and corruption-free organisations in Bangladesh (KII05). This organisation has established and well-recognised mechanisms to ensure accountability within their organisation, and with their partner organisations (KII05). A high-level PKSF official supports this:

---

[8] Deputy Commissioners are the administrative and revenue chiefs at district levels in Bangladesh.

> We're maintaining internal and external auditing systems, and we've contact with the donors if there are any discrepancies with the existing systems. In the case of climate change projects, WB is rating our project performances as satisfactory. There are four rating grade, satisfactory is the second grade and highly satisfactory is rarely observed. ADB, UNDP, FAO and others are rating us as satisfactory. So, we've established a globally recognised accountability system in PKSF. (KII05)

The majority of climate projects are based on adaptation, with less emphasis on mitigation. Some typical problems are evident in relation to project implementation. The main issues are the design and implementation of adaptation projects. For example, waste management of the Buriganga River should not be a BCCTF project, as this type of project will create significant questions about fund use. However, a few doubts remain around whether these are mainly related to climate change (KII24). Therefore, accountability is vital to make climate change-related projects more successful.

4.8 Governmental Organisation Transparency in Climate Change Activities

Transparency is a mechanism to ensure the accountability and legitimacy of governance systems (Biermann et al., 2010). Some reports exist regarding misuse of resources, especially of a financial nature, and fraudulent practices by both GOs and NGOs. In particular, newspaper reports focus on how funds from the BCCT have been misused. Moreover, corruption (as a major impediment to development) is distracting to many preparedness activities focused on minimising climate change effects (Bhuiyan, 2015; Younus, 2016). As a mid-level staff member of a reputed NGO explains:

> If you disclose more information relating to climate projects, there are little chances to manipulate or have corruption in the processes. So, it's important to disclose project information at every level, and need to incorporate locals in the consultation level. At the national level, it's very important to disseminate the project evaluation findings, so that various stakeholders could know about the project outcomes. At the project sites, there should be signboards mentioning information relating to the projects. Unfortunately, however, we didn't notice these in the majority of the projects and these indicate a breach of commitment. (KII26)

The GoB usually maintains control over fund and program management, which is another barrier to transparency (Khan, 2019).

There is a lack of transparency regarding funds allocation from the BCCT. BCCT officials have no specific criteria for the proper use of climate funds. They have failed to show transparency in managing climate funds (KII21). The following extracts provide some explanation of the transparency issue:

> As the government activities aren't so transparent, so people don't know about many decisions. (KII22)

> They (government departments) are transparent to themselves, but not to the public. How do they approve the projects? Why do they take the projects? How do they determine the

> funding level of the projects? Such information is not available to the public. We came to know these through our own processes of investigation. For example, in a particular project, BDT 10 million[9] was required for proper implementation but the government allocated only BDT two or three million. So, can you imagine what will be the outcomes of that project? (KII24)

> Donors' accountability remains relatively operational. When there is donors' funding, they ensure some degree of control over the funds. The donors' funded projects typically show a better degree of transparency compared to the purely government-funded projects. (KII25)

> We're trying to make the government more transparent and citizens more aware about climate finance and how we can monitor the progress of climate finance. Both of these will contribute to better governance on the one hand transparency from the government side and on the other hand accountability from people's side. (KII30)

PKSF is considered a transparent organisation by most, and is making significant effort in its use of climate funds (KII08) (KII30) (KII24). To maintain accountability and transparency in climate change-related programs, PKSF has undertaken initiatives, but these are not considered robust. Citizens are also coming forward to learn about the funding of climate projects (KII29).

4.9 Participation of Governmental Organisations in Climate Change-Related Activities

Good governance can ensure accountability, transparency and participation (KII21) (KII25) (KII28). Transparency, accountability, and participation always ensure the optimum use of resources (KII04). Climate change governance necessitates a participatory approach; that is, the participation of stakeholders at different levels (Sprain, 2016).

Participation depends upon the location where a project is taking place. If a project is implemented in an urban area, then the presence of more experts is observed. For example, for a rising sea level project in Dhaka, the DoE invited experts from CEGIS, IWM, IWFM-BUET, Bangladesh Rural Advancement Committee (BRAC), public and private universities and others outside academia. They invited relevant department representatives from the navy, coast guard, BIWTA, Bangladesh Water Development Board (BWDB), experts on water resources and representatives from other agencies. In contrast, for a similar study conducted in Khulna, a divisional city, could not gain the involvement of this number of experts. Invitations were issued to faculties from the University of Khulna and Khulna University of Engineering and Technology, agencies based in Khulna, and representatives of CSOs/NGOs. The following extracts explain the participation of relevant stakeholders in climate change-related activities:

> If I say, in terms of accountability and participation, we're doing unplanned tasks and corruption badly. (KII21)

---

[9] USD 1 is equal to 84.75 Bangladeshi taka (2021).

> If the project is consultant driven and the project director (PD) isn't very careful, then the various stakeholders' participation will not be satisfactory. In a DoE project, we try to involve our divisional officers. We ask them to invite relevant experts covering all the relevant sectors of the project. I've observed several workshops in the last few years which were well-represented. (KII03)

During the NAPA preparation, six divisional consultation meetings were conducted. BCCSAP was prepared through three consultation meetings and a few experts provided their views. In the case of NDC, no consultation meeting was held. Clearly, participation can be a significant consideration. That is why project results become government documents and not national ones (KII10). The following extract explains this:

> I don't see any ownership of these policies and documents because these don't reflect any of our suggestions, observations, demands and recommendations. So, how do we feel the ownership of these documents? (KII10)

BCCTF laws were developed in October 2009. The first meeting was held with 11 ministers and five secretaries, but there was no CSO or public participation in the approval of projects (KII21). Donor-funded projects follow various steps to ensure the participation of other stakeholders, whereas government-funded projects do not follow these practices (KII22). The following extract explains this:

> In Bangladesh, there is no prior general consultation by the GOs with the common people, it's not like a European country where government bodies will arrange meetings in the Town Hall. What GOs are saying in Bangladesh is considered as one kind of eye wash? In contrast, when GOs and NGOs are working together at the community level, they often consider the participation of the relevant stakeholders. (KII24)

From the evaluation or tracking of about 50 projects, participation was observed to be minimal, especially at the community level. The produced list of beneficiaries is considered faulty and the actual victims of climate change are absent (KII26).

It has been observed in a few cases that where people are engaged in project monitoring systems, the ultimate outcomes of those projects were much better. For example, cyclone shelter construction projects under BCCRF from 2012 to 2014 incorporated the local school management committee as an oversight agency. They were given some training to oversee the works, and their performance was praiseworthy. As an example, because of participants' reports, works were suspended until the contractor stopped using unsuitable materials. Community monitoring is important for effective implementation of climate change-related projects (KII26). Currently, NGOs and CSOs are coming forward and participating in various workshops, and are maintaining networks (KII29).

4.10 Effectiveness of Governmental Organisations in Climate Change-Related Activities

Government effectiveness is linked with the capacity, coordination and participation of stakeholders associated in climate change activities. Government officials often face difficulties moving along with climate projects because of limited capacity, a lack of coordination and an imbalance in participation. In particular, who is responsible for specific aspects of a project must be very clear. Government departments are using up to one-third of project funds allocated to climate change issues for enhancing staff capacity building, but local government officials experience difficulties with the lack of funds when trying to work directly on the project's climate change issues. Capacity development is not taking place as anticipated (KII10).

Levels of good governance are not very effective for CCA-related operations in Bangladesh (Bhuiyan, 2015). However, corruption and mismanagement in addressing climate change issues are diminishing the expected outcomes, because of governmental intervention (Ministry of Foreign Affairs of the Netherlands, 2018). A public university professor explains this situation:

> Actually, a good number of climate change-related projects are taking place in the country now. We expect effective results from these projects but we're not getting as much as we want. A major portion of the funds is misused due to various factors, such as project selection, fund allocation processes, political interferences, bureaucratic problems, corruption etc. So, these projects can't come up with the expected outcomes for the climate-vulnerable people. (KII27)

Climate projects adopted by government departments experience challenges such as not revealing the sources of funds in project implementation, the issue of political connections in contractor selection, the use of low-quality materials in construction works, and the absence of community participation in project implementation, supervision and monitoring (KII26) (TIB, 2013). A senior research fellow from a reputed national research institute explains this:

> We can't assess the effectiveness of climate projects because we do campaign more to focus on other issues. If we see long-term effectiveness on this aspect from 2005, we can see some improvements in policy level, but we can't see much improvement at the grassroots level because the government doesn't emphasise much on good governance issue in many climate projects. Though we're receiving funds from BCCT, there are still some debates concerning budget allocation. (KII24)

Government effectiveness issues are important, but this is not yet practised as a primary consideration in Bangladesh (KII29). Climate change issues have been under discussion for a time and government delegations have gone abroad to negotiate in various forums. Upon returning to Bangladesh after these negotiations, the scenarios have changed. Bangladesh is continuing to implement various projects and programs focusing on climate change at the present (KII08).

4.11 Funding Governmental Organisations for Climate Change Activities

Government-allocated climate funds are managed by ministerial committees, secretary committees and technical committees, and they are responsible for the approval of proposals and use of these funds (KII24). The Ministry of Finance is responsible for the allocation of funds. This ministry recently introduced a new system for climate fund management, the Climate Fiscal Framework. This program is supported by UNDP and the German Agency for International Cooperation (GIZ). This program seeks to keep records of climate fund use in different sectors (KII23).

The development budget receives only 10% from external sources and 90% from internal sources. Major development projects are fully funded by external sources. Projects based on external funds often address climate change issues, but officials involved in project planning processes raise questions relating to cost increases. The cost will usually increase as they consider the magnitude and intensity of natural disasters arising from climate change. If the government implements various projects using their own funds by addressing climate change issues then this would be internalised. In externally funded projects, donors place various conditions on integrating climate change issues such as disaster, environment, social and gender elements. Climate change is a focus in various major government projects. Government-funded projects are disbursed by BCCT to provide token money for climate projects. Through BCCT projects, the government was supposed to demonstrate its capacity, and such projects would be used as pilot projects or as learning opportunities. BCCRF's operation failed because of the lack of effective governance practices and the lack of required support from the WB (KII07). Bilateral donors are trying to invest funds in current climate projects (KII22). A senior research fellow from a reputed research organisation describes the situation:

> Actually, we always consider the climate funds as an important issue, and we've started to work for receiving more funds from different sources to address climate change issues. However, the government mechanisms aren't ready yet to make the best uses of climate funds. (KII24)

To meet the requirement of BCCSAP until 2015, an extra USD 10 billion will be required to proceed with regular development (KII21). The government has planned to spend 60% of funds for climate change from their own sources, whereas the rest (40%) will come from other organisations and donors. The government has taken on most of the responsibility for this; a good sign that this money will come from [government] revenue (KII10).

The Annual Development Programme (ADP) of Bangladesh is worth more than $18 billion. This has increased by an annual average of 21% over the last six years. The ADP Bangladesh is higher than the total national budget of 49 LDCs. The ADP has increased over the time (as explained above), but the BCCSAP or other plans did not lead to this rise in the amount of ADP (KII21) (KII29). The following extract explains this:

> If we see the ADP, the utilisation rates are not at a satisfactory level over the years. Various ADP projects don't focus the climate change issue properly. Few projects emphasise the climate change issue but the results are not expected level. So we can't exploit these benefits due to lack of proper fund management systems but these investment outcomes will last for another 50 years. (KII21) (KII29)

Moreover, climate change-related projects need additional money (KII21) (KII22) (KII24). For example, to develop an embankment to tackle rising sea levels, a certain amount of money is required, but when climate change is involved, more money will be needed. The finance ministry usually does not allocate additional funds for climate change issues in development projects if these are not included at the planning stage (KII21). A university professor explains this:

> Bangladesh always needs additional money and these have to come from external sources. I don't see any problem with this. If you're smart enough, and if you know the diplomacy and the art of resources mobilisation, that is a strength. We'd still continue to mobilise the foreign resources, given the fact that we'll face difficulties to have enough funds to support a wide range of development programs with our own resources. If we can mobilise competitive foreign funds in climate projects, then it would be a good indicator of project efficacy. (KII25)

Bangladesh has expended a great amount of funding on climate change issues over many years. When good governance is absent and corruption is ubiquitous, these funds cannot be used properly (KII26) (KII28). To implement large projects, locally available funds are not enough. The government therefore needs external funds to make these happen. Local funds can usually only support managing climate change effects on local livelihoods (KII28). An expert on climate change explains this:

> The sustainability of climate change-related programs becomes a major challenge when the funding is over. We're also experiencing a general decline in funding of development projects with the official announcement of the graduation from a LDC to a developing country status. If Bangladesh is no longer qualified for the special funding that we're getting on a routine basis, the funding for development projects, including climate change issues, will go down. We're observing these issues in the last few years. Although the government is claiming that they have substantial funding and that they can go forward with their own resources, this is actually merely political rhetoric in this country. (KII22)

Climate expenditure is significant to those departments or NGOs to whom the subsidy is given, regardless of whether these funds will reach vulnerable people, or whether they are well targeted. There are several entry points in the budgeting system. First, climate change issues must be incorporated in macro-economic modelling and planning, and it should then cascade down through the medium-term budget framework (MTBF). It will then be integrated into ministerial meetings where the issue can be addressed. Second, the question of whether people are gaining benefits from climate investments and whether such investments meet the expected demands of vulnerable people needs to be asked and assessed. Otherwise, fiduciary performances cannot also be assessed (KII29). The following extracts explain this:

> To oversee the issues we're trying to follow the audit protocol of the CAG office. Oversight of the parliament is ensured through the committee named 'Public Accounts Committee'

> which acts upon audit reports submitted by the CAG. If we put climate change issues in the audit report, then the parliamentarians will be informed about the issues, and they can put some recommendations to them. Thus the fiduciary risk will be reduced. (KII29)

> We see, in reality, that GOs search the funding windows and then go for the development of the projects rather than developing the project first. This compromises the quality of the project. So, they're focused on short-term funding availability rather than long-term sustainability. (KII26)

With PKSF, funds are mainly sourced internally, and sometimes the government provides limited grants or loans. For example, the government allocated BDT 250 crores (USD 29 million) to PKSF as loans in 2018 (KII05). They also implemented programs using their own resources in the previous few years (KII05). The following extract explains this situation:

> Sometime we're getting funds from WB, Asian Development Bank [ADB], FAO, UN organisations, DFID, GIZ etc. But our funding isn't sufficient to cover various activities particularly for climate change adaptation. Up to 2013, we've received only BDT 25 crores (around USD 3 million), but after that we didn't receive any funds, and now it is completely stopped. (KII05)

Three projects have been funded under climate change conventions: (1) LDC funds; (2) the Special Climate Change Fund (SCCF) for developing countries where funding is limited; and (3) the GCF, where ERD acts as the NDA in Bangladesh. ERD is trying to obtain access to more funds from GCF, they provide a good level of funding. Unfortunately, these funds do not provide enough resources to meet project needs. The following extracts explain this dilemma:

> We're getting prepared for the good projects from 2020 and onwards. We're expecting to receive more funds. We're trying to take preparation, but we need to improve at various levels such as planning, implementation, training and procurement. This will support us to get more funds and implement these projects efficiently. Once we receive funds and cannot implement properly then it'll be difficult to get funds in future. (KII03)

> We've accessed already USD 26/27 million and we'll be able to access USD 13/14 million. Various sources are GCF, bilateral funds and multilateral funds for LDCs with low interest rates, and so we can avail these loans until 2024. We need to be strategic to declare as a developing country otherwise we'll lose many facilities. When we (GOs and NGOs) prepare proposals for climate funds, we need to put our status as LDCs. (KII04)

Bangladesh is facing difficulties accessing various external funds such GCF, adaptation funds, and the LDC fund (KII22). Questions arise regarding how the funds are allocated for climate change projects. Different ministries submit projects to the Ministry of Planning. Ministries that submit better proposals usually receive more funds. In the Seventh Five-Year Plan, some priority issues are highlighted and some special allocations are given for those purposes (KII04).

Sectoral allocations are important to maintain a balance of work on climate change across various areas. Allocations of funds are determined based on the project proposals received. If the government initiates the sectoral allocation of funds, then the concerned departments or ministries prepare proposals with support from experts. This may be one way of fund allocation at the government level. Conversely, if a particular community faces climate vulnerabilities, they deserve climate change funds. The concerned government departments of that particular area should then submit project proposals to gain funding to implement climate projects for that community (KII04).

Therefore, various windows are open for the government to obtain climate funds. The main challenges are that government bodies need to be prepared to access those funds and ensure the best use of those funds. Although the government has developed funding policies on climate change projects, these polices need to focus on proper use of funds.

4.12 Changes in Climate Change Governance

Many changes have occurred relating to climate change governance issues in Bangladesh over the years (KII24) (KII25) (KII26) (KII27) (KII29). Climate change governance systems started to evolve in Bangladesh after the Copenhagen Accord 2009. Change was accelerated by the involvement of several developed and developing countries and the opening of various funding windows. Global initiatives on climate change issues have had some influence on local climate change governance practices. The Bangladesh government has focused on developing plans and programs relating to climate change, with the involvement of major institutions. However, climate change governance did not gain much momentum as it expected; rather, it declined in terms of effectiveness (KII27). A mid-level government official explains this situation:

> We, government organisations, started to undertake mitigation policies with the developed countries, but couldn't precede much due to huge cost. Then the government commenced with adaptation policies which are both good and bad. Good in the sense that we've to adapt it and adaptation has no limitations. We've to adapt climate change issues in various sectors, but the adaptation practices are risky as well. Like mitigation, adaptation requires a huge amount of money, technologies and professionals also, and Bangladesh can't afford the large scale of adaptation due to its own limitations. (KII27)

When Bangladesh began working on climate change issues the government did not have fundamental goals to guide the country into effective climate governance. Instead, they continued their activities by considering only pressing issues. At the same time, they planned to enhance the capacity for accessing climate funds. For example, in the past, no government entities could submit an application for GCF because of a lack of internal governance controls. Although climate vulnerability is increasing, various government departments continue to follow standard procedures. This has meant that governance systems have not improved at the expected level. After the Paris Agreement 2015, climate change was endorsed as an issue in parliament and this became legally binding at the national level (KII27).

Bangladesh is trying to work with other partner organisations on climate finance to improve governance systems. This is termed the Climate Finance Transparency Mechanism (CFTM). Stakeholders are trying to make governance systems more transparent, and are encouraging citizens to be more aware of climate finance and monitoring issues. Both issues will contribute to developing better climate governance, with government transparency on one side and accountability from the people on the other (KII30).

The MoEFCC updated the Environment Policy 1992 and incorporated climate change issues. The Eighth Five Year Plan (2020–2025) also integrated climate change issues. These are significant improvements in the sector. The following extract explains this progress:

> Our ministry has taken various measures to enhance the capacity of the existing manpower to address the climate change issue which was absent a few years back. We're trying to increase the number of staff to advance the present situation. If the government approves the additional manpower, we can conduct some innovative tasks in future apart from the regular operations. (KII03)

Climate change governance is evolving in Bangladesh. The government is taking various measures for undertaking climate change-related projects. The environment ministry was renamed recently to indicate a new emphasis on climate change issues (KII24). The following extract demonstrates these changes:

> Climate change governance practices are changing over time at the government level. A good set of policies, along with the regulatory bodies are in place. Government organisations are prioritising and improving the tools, techniques and the processes for climate change-related programs. Government officials are gathering experiences through various activities, and they are performing better nowadays. If good governance is present then many issues could be overcome. (KII22)

BCCT started in 2010. Initially it was launched as a project that was completed in 2012; in 2013, it was transformed into a trust. Many staffs from that BCCT project were merged into the BCCT. A trustee board also operated in the early states. A mid-level official of BCCT explains the evolving structure:

> A checklist is incorporated in BCCT's activities to ensure monitoring in 2015. Eventually transparency and accountability are emphasised more. So, it's an important initiative to ensure transparency in the BCCT's work. The number of submitted project proposals has increased over the years, and that's why the present institutional capacity needs to be upgraded. We've forwarded a proposal for 123 more staff for BCCT to accelerate our activities. (KII02)

Delta Plan 2100 is a major breakthrough in terms of climate change recognition. If it can be managed properly, it will bring considerable achievements to the sector, and will take Bangladesh a long way (KII29). An expert on climate change issue offers a different perspective on climate change governance:

> I don't find any significant changes in climate change governance over the years. Many successes will be highlighted on climate change issues, but these will not bring any positive changes in the expected time through which we can reach our destination. If we move like a slow moving tortoise, then climate change will affect us badly, and we'll not be able to reach it. (KII21)

The Ministry of Finance is updating the Climate Fiscal Framework and is planning to reform and instigate innovative measures to promote climate change issues, such as enforcing taxes for polluting industries. The government is planning to ban reconditioned cars and those that pollute the environment. Three-wheel auto rickshaws, which were in the streets 20 years ago, are now banned by the government. Bangladesh Bank (BB) is now offering loans through a GCF, which are interest-free or at very low interest rates. If industries are developed with minimum emissions, they can access these bank loans. Such climate friendly initiatives have appeared in many sectors over the last ten years in Bangladesh (KII23).

Dhaka South City Corporations has declared that house owners will be rewarded with a 10% tax rebate if they contribute to a greener city through rooftop gardens, balcony gardens or gardens in compounds (The Daily Star, 2016). For example, PKSF started work on climate change issues in 2012. They intended to work on a result-based monitoring system from 2015, and this is gradually improving. The internal monitoring systems have been upgraded in the last three or four years and these systems are now applicable to other relevant activities, including climate change issues (KII05).

Therefore, GOs are at work on various aspects to address climate change issues and they are collaborating with other sectors simultaneously. Climate change governance is addressing its barriers, and is becoming more functional in response to changing scenarios.

4.13 Discussion and Conclusion

The GoB has undertaken several measures relating to climate change issues. Various GOs are focusing on climate change. They have implemented a large number of climate change-related projects already and more are in train. They have established an effective set of policies, protocols and plans, and at least some active institutions are in the right place to move forward. Bangladesh is putting considerable effort into ensuring the rights of LDCs and developing countries regarding climate change in the international arena. The Bangladesh government signed the Paris Agreement, and has also ratified it in parliament.

Government departments are emphasising climate change issues to ensure improved results from climate projects. Accountability systems at the government level are evolving, and government departments are focusing on this issue to reduce corruption. Monitoring tools are being developed and applied to realise the expected outcomes of climate change-related projects.

What needs to be emphasised, however, is that the government's implementation of effective monitoring, transparency and fiscal controls has not yet reached the acceptable levels needed for good governance. This relates to all of development projects, including climate projects. Transparency is an essential mechanism in this sector. Without maintaining transparency in climate projects, government initiatives cannot achieve their targeted goals, and the most climate-vulnerable people will not see the benefits they expect.

GOs are working on climate change issues from their relevant positions. Their combined efforts are not visible in a way that reveals effectiveness at the community level. Moreover, a lack of coordination among stakeholders is evident in many climate projects. For example, the BCCSAP and the BCCT did not specify the role of local government, and there was no reflection of the BCCSAP in the local government's plans, strategies, activities or budgets.

The institutional mechanisms are in place to promote climate change-related activities, but there are few controls and limited oversight linked with these mechanisms. These mechanisms are not successful owing to the lack of combined effort from relevant departments at various implementation levels. The robustness of MoEFCC is needed to achieve better outcomes from current institutional mechanisms. The efficiency of officials from the relevant ministries is not at a level where it can strengthen climate project implementation processes. Moreover, climate change governance is not well understood by the majority of government officials. Generally, they cannot provide the required inputs for the different stages of climate project implementation.

The Bangladesh government has developed its own funding department to mobilise climate funds more efficiently. These funds are, however, not enough to enable full implementation of climate projects. It is a big challenge for the government to mobilise enough funds for climate change-related activities. So, the government often seeks support from the donors and other international organisations to speed up their efforts on climate change issues.

The government is working on various climate change initiatives at the national and international level, but there is a lack of political commitment in climate governance practices. Despite the introduction of various measures to eliminate corruption and malpractice, the government system is not prepared enough to achieve the expected level of accountability, transparency, participation and government effectiveness. To date, effective climate change governance has not progressed to a point where it can overcome climate change effects and make people more resilient. After all, GOs are responsible for the overall management of climate change governance practices, and effective climate change governance can enhance national resilience.

Chapter 5 outlines the climate change governance practices of NGOs in Bangladesh.

References

Ahamed, N., Assaduzzaman, M., Bhuyan, M. H. R., Iqbal, Z., Parvin, M., & Asaduzzaman, M. (2018). *Assess the monitoring, evaluation procedure of BCCT and the impact assessment of CCTF projects*. Dhaka, Bangladesh. http://bidslink.bids.org.bd/bidsorgbd/completed_research/Executive%20Summary_CCTF%20Final%20Report.pdf

Ahmed, A. U. (2006). *Bangladesh climate change impacts and vulnerability: A synthesis*. Comprehensive Disaster Management Programme, Government of the People's Republic of Bangladesh. https://www.preventionweb.net/files/574_10370.pdf

Alam, M., & Rabbani, M. D. G. (2007). Vulnerabilities and responses to climate change for Dhaka. *Environment and Urbanization, 19*(1), 81–97. https://doi.org/10.1177/0956247807076911

Alam, K., Shamsuddoha, M., Tanner, T., Sultana, M., Huq, M. J., & Kabir, S. S. (2011). The political economy of climate resilient development planning in Bangladesh. *Institute of Development Studies Bulletin, 42*(3), 52–60.

Ayers, J., Huq, S., Wright, H., Faisal, A. M., & Hussain, S. T. (2014). Mainstreaming climate change adaptation into development in Bangladesh. *Climate and Development, 6*(4), 293–305. https://doi.org/10.1080/17565529.2014.977761

Bangladesh Climate Change Trust. (2017, January–March). Newsletter: To enhance the capacity to build climate resilient Bangladesh. *News Letter*, p. 1.

Begum, S. A., Uddun, A. M. K., Ernst, R., & Haque, N. (2006). *Climate change and Bangladesh: Annotated bibliography.* http://www.climatechange.gov.bd/sites/default/files/Bibliography%20(Annotated)%20on%20Climate%20Change%20and%20Bangladesh.pdf

Bhuiyan, S. (2015). Adapting to climate change in Bangladesh. *South Asia Research, 35*(3), 349–367. https://doi.org/10.1177/0262728015598702

Biermann, F., Betsill, M. M., Gupta, J., Kanie, N., Lebel, L., Liverman, D., Schroeder, H., Siebenhüner, B., & Zondervan, R. (2010). Earth system governance: A research framework. *International Environmental Agreements: Politics, Law and Economics, 10*(4), 277–298.

Choudhury, G. A., Catharien Terwisscha van Scheltinga, C. T. V., Bergh, D. V. D., Chowdhury, F., Heer, J. V., Hossain, M., & Karim, Z. (2012). *Preparations for the Bangladesh Delta Plan.* info.alterra@wur.nl

Finance Division. (2018). *Climate finance for sustainable development.* www.mof.gov.bd

Firoz, R. (2018, February 16). An analysis of the BCCSAP projects implemented in Bangladesh. *Dhaka Tribune.* https://www.dhakatribune.com/tribune-supplements/tribune-climate/2018/02/16/analysis-bccsap-projects-implemented-bangladesh

GoB. (2018). *Climate financing for sustainable development: Budget report 2018–19.* Dhaka, Bangladesh. www.mof.gov.bd

GoB. (2020). *Eighth Five-Year Plan July 2020–June 2025: Promoting prosperity and fostering inclusiveness.* General Economics Division (GED), Bangladesh Planning Commission, Government of the People's Republic of Bangladesh. http://www.plancomm.gov.bd/

Government of Bangladesh. (2005). *National adaptation programme of action (NAPA).* https://unfccc.int/resource/docs/napa/ban01.pdf

Government of Bangladesh. (2009). *Bangladesh climate change strategy and action plan 2009.* Dhaka, Bangladesh. http://www.moef.gov.bd/site/page/97b0ae61-b74e-421b-9cae-f119f3913b5b/BCCSAP-2009

Huq, S. (2017, April 16). Climate finance in Bangladesh. *Daily Star.* http://www.thedailystar.net/op-ed/politics/climate-finance-bangladesh-1209307

Huq, S. (2020, July 15). Climate change: Turning Bangladesh into the learning capital of the world. *Daily Star.* https://www.thedailystar.net/opinion/politics-climate-change/news/climate-change-turning-bangladesh-the-learning-capital-the-world-1930405

Huq, S., & Rabbani, G. (2011). Climate change and Bangladesh: Policy and institutional development to reduce vulnerability. *Journal of Bangladesh Studies, 13*(1), 1–10.

Implementation Monitoring and Evaluation Division. (2008). *Strategic plan 2008 to 2013.* https://imed.gov.bd/site/charter_of_duties/1f8bbaf5-0b58-4363-a944-b2dc137cd2e1/Strategic-Plan

Irfanullah, H. M. (2020, February 18). Biodiversity and climate change: Our missed opportunities. *Daily Star.* https://www.thedailystar.net/opinion/politics-climate-change/news/our-missed-opportunities-1869364

Khan, A. A. (2019). Social and legal barriers to improving human rights of climate change displaced people in Bangladesh. *Journal of Interrupted Studies, 2*(1), 103–117. https://doi.org/10.1163/25430149-00201004

Khatun, F. (2020, December 21). Expectations from the Eighth Five Year Plan. *The Daily Star.*

Lopa, F. G. R., & Ahmad, M. M. (2016). Participation of CSOs/NGOs in Bangladeshi climate change policy formulation: Co-operation or co-optation? *Development in Practice, 26*(6), 781–793. https://doi.org/10.1080/09614524.2016.1200536

Luxbacher, K. (2011). *Inside stories on climate compatible development*. http://unhabitat.org.ir/wp-content/uploads/2019/04/Bangladesh-InsideStory_5pp_pr4F_LR1.pdf

Mills-Novoa, M., & Liverman, D. M. (2019). Nationally determined contributions: Material climate commitments and discursive positioning in the NDCs. *Wiley Interdisciplinary Reviews: Climate Change*. https://doi.org/10.1002/wcc.589

Ministry of Environment, Forest and Climate Change [MoEFCC]. (2005). *National adaptation programme of action*. MoEFCC, Government of Bangladesh. https://www4.unfccc.int/sites/NAPC/Country%20Documents/Parties/ban01.pdf

Ministry of Environment, Forest and Climate Change. (2009). *Bangladesh climate change strategy and action plan 2009*. MoEFCC, Government of the People's Republic of Bangladesh.

Ministry of Foreign Affairs of the Netherlands. (2018). *Climate change profile Bangladesh*. www.government.nl/foreign-policy-evaluations

Pervin, M. (2013). *Mainstreaming climate change resilience into development planning in Bangladesh*. http://pubs.iied.org/10045IIED

Rahman, M. M., & Huang, D. (2019). Climate governance initiatives: Snapshots from Bangladesh. *Journal of Geoscience and Environment Protection, 7*(9), 131–147. https://doi.org/10.4236/gep.2019.79010

Rai, N., Huq, S., & Huq, M. J. (2014). Climate resilient planning in Bangladesh: A review of progress and early experiences of moving from planning to implementation. *Development in Practice, 24*(4), 527–543. https://doi.org/10.1080/09614524.2014.908822

Sarker, M. A. (2019). *Bangladesh Delta Plan 2100 (Bangladesh in 21st century)*. https://www.gwp.org/globalassets/global/gwp-sas_images/gwp-sas-in-action/ldai/bdp-2100-ppt.pdf

Siddique, A. B. (2016, April 5). Donors shut down climate fund, Bangladesh to lose $50m. *Dhaka Tribune*. www.dhakatribune.com/bangladesh/2016/apr/05/donors-shut-down-climate-fund-bangladesh-lose-50-m

Sprain, L. (2016). Paradoxes of public participation in climate change governance. *Good Society, 25*(1), 62–80.

Stock, R., Vij, S., & Ishtiaque, A. (2020). Powering and puzzling: Climate change adaptation policies in Bangladesh and India. *Environment, Development and Sustainability, 23*, 2314–2336. https://doi.org/10.1007/s10668-020-00676-3

The Daily Star. (2016). DSCC to reward rooftop gardening. *The Daily Star*, Dhaka. https://www.thedailystar.net/city/dscc-reward-rooftop-gardening-thru-tax-rebate-1261096

The Daily Star. (2020). Mobilise global support to address climate refugee issue: PM to G20. *The Daily Star*, Dhaka. https://www.thedailystar.net/country/news/mobilise-global-support-address-climate-refugee-issue-pm-g20-1961813

Thomalla, F., Cannon, T., Huq, S., Klein, R. J. T., & Schaerer, C. (2005). *Mainstreaming adaptation to climate change in coastal Bangladesh by building civil society alliances*. Paper presented at the Solutions to Coastal Disasters 2005, Charleston, USA.

Transparency International Bangladesh. (2013). *Climate finance in Bangladesh: Governance challenges and way out (executive summary)*. https://www.ti-bangladesh.org/beta3/images/max_file/es_cfg_executive_summary_english.pdf

Uddin, M. J. (2018). Destruction of river systems by using climate funds [Bengali]. http://www.prothomalo.com/bangladesh/article/1462686/

Vij, S., Biesbroek, R., Groot, A., & Termeer, K. (2018). Changing climate policy paradigms in Bangladesh and Nepal. *Environmental Science & Policy, 81*, 77–85. https://doi.org/10.1016/j.envsci.2017.12.010

World Bank. (2012). *Bangladesh climate change resilience fund*. https://www.worldbank.org/en/news/feature/2012/05/22/bangladesh-climate-change-resilience-fund-bccrf

Younus, M. A. F. (2016). Adapting to climate change in the coastal regions of Bangladesh: Proposal for the formation of community-based adaptation committees. *Environmental Hazards, 16*(1), 21–49. https://doi.org/10.1080/17477891.2016.1211984

Zamudio, A. N., & Parry, J.-E. (2016). *Review of current and planned adaptation action in Bangladesh.* https://www.iisd.org/system/files/publications/idl-55862-bangladesh.pdf

Chapter 5
Climate Change Governance Practices of Non-governmental Organisations in Bangladesh

5.1 Introduction

This chapter illustrates the climate change governance approaches employed by NGOs in Bangladesh. This chapter employed a systematic interview schedule to elicit precise responses to the subsequent study questions:

- What are the functions of NGOs in the development and execution of diverse climate change-related projects and programs?
- What measures do NGOs employ to address climate change governance?

This chapter initially examines the functions of NGOs in activities related to climate change. It subsequently delineates the climate change governance practices of NGOs, encompassing their institutional frameworks, accountability measures, transparency, stakeholder engagement, operational efficacy, resource reliance, and funding mobilization tactics. It subsequently evaluates the achievements and obstacles faced by NGOs in the management of climate change-related initiatives.

5.2 Non-governmental Organisations in Climate Change-Related Activities

NGOs are significant participants in the combination of climate change adaptation (CCA) and disaster risk reduction (DRR) (IPCC, 2012). At present, the efforts of local NGOs, in collaboration with donors and international NGOs (INGOs), concentrate on climate change adaptation (CCA) and disaster preparedness (Rahman & Huang, 2019). NGOs likewise concentrate on environmental concerns and have endeavored to cultivate, to some extent, equity and parity among nations and groups (Raustiala, 2001). NGOs, like to other stakeholders, endeavor to mitigate insecurity and safeguard individuals who may be susceptible to climate change (Bannerman et al., 2011).

Md. F. Hossain and S. Momtaz, *Evaluating Climate Change Governance in Bangladesh*,
https://doi.org/10.1007/978-3-031-98486-0_5

NGOs have established connections between authorities and local implementing agencies (Young & Dhada, 2013). A large number of extremely proficient NGOs function in Bangladesh. By 2020, the Bangladesh government granted approval to a total of 2501 NGOs (NGOAB, 2020). Currently, some 300 NGOs are engaged in disaster risk reduction and climate change initiatives (Rahman & Huang, 2019).

Bangladesh has one of the greatest densities of NGOs globally, and the government has established numerous regulations for their functioning. NGOs serve as a catalyst for development and make substantial contributions to numerous important Sustainable Development Goals (SDGs). The significant advancements of Bangladesh's NGOs frequently motivate other nations to foster active NGO involvement in development initiatives, particularly regarding climate change matters. In Bangladesh, NGOs have attained political and social legitimacy by serving as crucial collaborators in community development initiatives (Ilyas, 2010).

The operations of Bangladeshi NGOs are regarded as some of the most dynamic globally (Zafar Ullah et al., 2006). The NGO sector in Bangladesh is systematically structured and comparatively reliable (Gauri & Galef, 2005). The presence of NGOs in underdeveloped nations like Bangladesh exemplifies effective civic engagement and fosters good governance and efficient policy execution in the absence of strong national infrastructures (Rahman, 2016). Over the past three decades, the NGO sector in Bangladesh has transitioned from social mobilization to service delivery activities (Kabeer et al., 2012; Rahman, 2016). NGOs in Bangladesh have established an extensive network of service initiatives aimed at enhancing the lives and livelihoods of individuals vulnerable to climate change (Thomalla et al., 2005).

Numerous NGOs possess their own CCA strategies and include climate change concerns into their comprehensive operational domains. Opportunities exist to amend existing programs through the involvement of pertinent stakeholders, including NGOs. NGOs actively contribute to the comprehensive development of the nation (Baroi & Panday, 2018; Huda, 2020). They function in inaccessible regions, concentrating on disadvantaged and at-risk populations (Young & Dhada, 2013).

NGOs function in 19 coastal districts of Bangladesh (see Fig. 5.1), focusing on enhancing the lives and livelihoods of impoverished coastal populations (Ahmad & Rahman, 2011), and the current study was undertaken in one of these districts. Their initiatives encompass education, healthcare services, agricultural development, water and sanitation, microfinance, legal assistance, disaster risk reduction, and projects addressing climate change concerns. Non-governmental organizations are actively organizing volunteer groups for disaster response, focusing on disseminating information to mitigate vulnerability, constructing and managing cyclone shelters, supporting alternative livelihoods, and developing adaptation strategies to address the detrimental effects of such disasters. They are imparting knowledge to optimize the utilization of accessible coastal resources and assisting coastal communities in addressing climate change-induced calamities (Ahmad & Rahman, 2011).

NGOs engage in climate change adaptation at the community level; nevertheless, these initiatives are predominantly small-scale or pilot projects (KII23). For instance, several prominent NGO initiatives include approximately 100,000 of the estimated 32.5 million families in Bangladesh (KII21). Enhancing the efficacy of

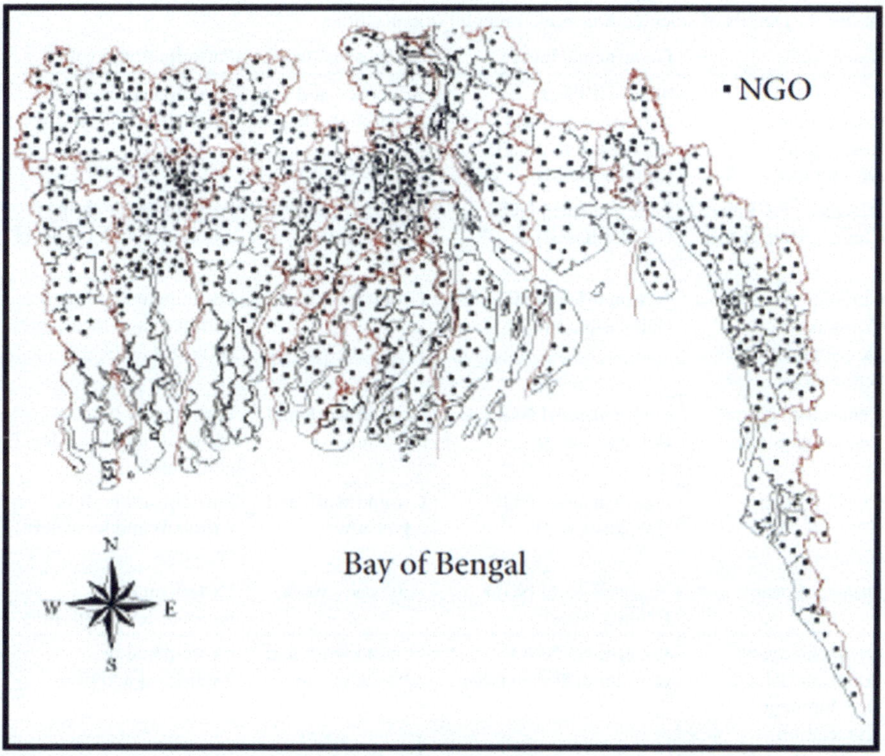

Fig. 5.1 Map of non-governmental organisations operating in 19 coastal districts. *Source* Program Development Office for Integrated Coastal Zone Development Plan (2005)

NGO programs could involve more efficient integration with government initiatives, notwithstanding the frequent misalignment between NGO activities and governmental operations. Climate change concerns are frequently associated with poverty, macroeconomics, and infrastructure development. Numerous disaster risk reduction efforts remain unaddressed by non-governmental organizations. Nonetheless, NGOs are excelling and offering credible novel models for the government to contemplate in the future.

The subsequent NGOs in Bangladesh were questioned to gather their perspectives on climate change governance practices within their organizations. All surveyed NGOs engaged in initiatives or activities that addressed climate change issues (see Table 5.1).

The NGOs chosen for this study tackle various facets of climate change (see Table 5.2). Their initiatives encompass climate change education, innovation in climate solutions, technological demonstration and dissemination, capacity development, and the enhancement of community resilience. They additionally promote changes at the policy level. Consequently, NGOs assist governmental agencies in formulating strategies and policies pertaining to climate change matters. NGOs

Table 5.1 Details of selected non-governmental organisations

Name	Governance level	Category of work	Participation level
Centre for Participatory Research and Development	National-level non-governmental organisation (NGO) (Dhaka-based)	Advocacy and research work	Participated in national-level activities
Network on Climate Change, Bangladesh	National-level NGO (Dhaka-based)	Advocacy and ground work	Participated both in national- and local-level activities
Christian Commission for Development in Bangladesh	National-level NGO (Dhaka-based)	Ground work and advocacy	Participated both in national- and local-level activities
Centre for Natural Resources Studies	National-level NGO (Dhaka-based)	Ground work and advocacy	Participated both in national- and local-level activities
COAST Trust	National-level NGO (Dhaka-based)	Ground work and advocacy	Participated both in national- and local-level activities
Disaster Forum	National-level NGO (Dhaka-based)	Advocacy work	Participated in national-level activities
Dwip Unnayan Songstha (Island Development Association)	Local-level NGO (coastal district-based)	Ground work and advocacy	Participated in local-level activities

Source Hossain and Momtaz (2024) and Lopa and Ahmad (2016)

endeavor to guarantee the rights and involvement of populations susceptible to climate change during the planning and execution stages. Additionally, NGOs at the local level endeavor to fortify local institutions and enhance public understanding of climate change to advance sustainable development projects. Moreover, NGOs are conducting active research to uncover policy deficiencies and community-level requirements.

The surveyed NGOs possess substantial expertise in disaster risk reduction and climate change (see Table 5.3). Since 1982, Dwip Unnayan Songsthau (DUS) has conducted operations centered on disaster events in coastal sub-districts. DUS addresses water and sanitation challenges, livestock management, disaster risk reduction, and microfinance. They have also implemented climate-resilient crop species in their operational regions (KII13). The CCDB has been operational since 1996 and is well-regarded for its climate change activities (KII12). CCDB commenced its efforts in relief and rehabilitation in 1996, subsequently transitioning to address climate change. CCDB engages in multiple facets of climate change, including knowledge generation and dissemination, training providing, advancements such as

Table 5.2 Roles of surveyed non-governmental organisations in climate change-related issues

Name of NGOs	Main areas of activities focusing on climate change
Centre for Participatory Research and Development	Supporting government organisations directly for developing various plans and policies on climate change Communicating findings from the grassroots to the government level Awareness-raising programs on climate change issues Advocacy and policy dialogue Research on climate change issues
Network on Climate Change, Bangladesh	Implementing programs to reflect the voice of the most vulnerable groups in climate change-related policies and plans at the national and global levels Working on the rights, participation and transparency of the most vulnerable people Working on climate change advocacy
Christian Commission for Development in Bangladesh	Working to develop a climate sensitive and resilient society Working on knowledge generation and sharing, climate learning and education, innovation of climate solutions, technology demonstration and dissemination, capacity development and community resilience building
Centre for Natural Resources Studies	Climate change adaptation and mitigation, and disaster planning Working on natural resource management and natural resource systems affected by climate change
COAST Trust	Funding disaster risk reduction Awareness programs to make people more resilient to climate-related shocks Advocacy to make policy-level changes Climate finance Work in the coastal communities affected by climate change events
Disaster Forum	Working on disaster risk reduction issues with special focus on preparedness Working on environmental issues Promoting the rights of all vulnerable communities Promoting alternative perspectives on disaster mitigation

(continued)

Table 5.2 (continued)

Name of NGOs	Main areas of activities focusing on climate change
Dwip Unnayan Songstha (Island Development Association)	Enabling coastal community capacity to cope with natural disasters Raising awareness of climate change adaptation in coastal communities Protecting coastal ecosystem and promoting eco-friendly agriculture and social forestry Strengthening local institutions to promote development initiatives

Source Hossain and Momtaz (2024)

a climate technology park, and community capacity enhancement (KII12). Additionally, CCDB participates in international climate change negotiations and advocates for climate-resilient construction methods at both national and local levels (KII12).

The Centre for Natural Resources Studies (CNRS) commenced operations in 1994 and is one of the non-governmental organizations in Bangladesh addressing environmental concerns (CNRS, 2020). This NGO engages in the execution of diverse initiatives related to livelihoods, advocacy and governance, disaster risk reduction, climate change adaptation, gender and development, and microfinance (KII09). DF, a nationwide disaster preparedness network located in Dhaka, has been addressing catastrophe and vulnerability issues since 1994 (DF, 2020). This NGO strives to guarantee accountability within development agencies and to advocate for the rights of catastrophe victims (KII11).

The NCCB commenced operations in 2005. It executes initiatives centered on disasters and climate change (KII10). COAST Trust and the CPRD both commenced operations in 2007 (KII07, KII08). COAST Trust has engaged in climate change-related policy modifications since its establishment. It addresses multiple climate-related concerns, including the construction of embankments, the reformation of the Water Development Board, fisheries, forestry, and climate-adaptive technology

Table 5.3 Work experience of surveyed non-governmental organisations

Name of non-governmental organisation	Start year
Centre for Participatory Research and Development	2007
Network on Climate Change, Bangladesh	2005
Christian Commission for Development in Bangladesh	1996
Centre for Natural Resources Studies	1994
COAST Trust	2007
Disaster Forum	1994
Dwip Unnayan Songstha (Island Development Association)	1982

Source Hossain and Momtaz (2024)

(KII07). CPRD engages in initiatives concerning disaster risk reduction, climate change, biodiversity, urban resilience, displacement and migration, and renewable energy (CPRD, 2020). Many NGOs launched their operations by concentrating on climate change concerns following significant disasters.

5.3 Non-governmental Organisation Institutional Arrangements

Institutional arrangements pertain to the preparedness of organizations, based on their current structure, to address climate change challenges. NGOs operate according to their institutional frameworks, which facilitate a more systematic execution of their activities. Numerous NGOs formulate regulations and embrace informal standards for their routine operations, which subsequently impact other stakeholders (Allard & Martinez, 2008). The governance practices of NGOs on climate change may exhibit greater transparency and accountability if comprehensive approaches were adopted by governmental organizations, coupled with a more formalized participation of NGOs in decision-making processes (Biermann & Gupta, 2011).

The CCDB functions in compliance with its constitution, as sanctioned by the Joint Stock Company of Bangladesh (RJSC) and NGOAB. A commission was established under this constitution, with members fulfilling a distinct role in decision-making processes (see Fig. 5.2). Commission members do not actively serve CCDB; nonetheless, they oversee ongoing activities, financial strategies, and various other matters. The executive director is accountable for the organization's operational matters and is authorized to make decisions pertinent to their domain of responsibility. CCDB administers four thematic programs: comprehensive poverty alleviation, climate change mitigation, resettlement, and microfinance. The climate change initiative possesses a distinct organizational structure led by a program director. Within this program, contingent upon the team's requirements, one to four program managers receive support from personnel on temporary assignments. The quantity of staff is contingent upon program levels. CCDB is systematically organized, upholds explicit norms and regulations, and possesses strong governance frameworks. The organization first encountered internal difficulties, which was resolved effectively. CCDB is regarded as having a commendable reputation concerning team makeup, age, and the execution of several activities.

CNRS is administered and regulated in accordance with the provisions outlined in its constitution. It is governed by two monitoring entities: a general board consisting of 21 members and an executive committee of seven members. The executive committee, in conjunction with directors, advisers, and staff members, is accountable for the daily management and functioning of the organization. CNRS possesses operational manuals, policies, and instruments that constitute the foundation and guiding principles for governance and operational affairs. Other divisions of the NGO manage the governance of organizational affairs, including the execution of

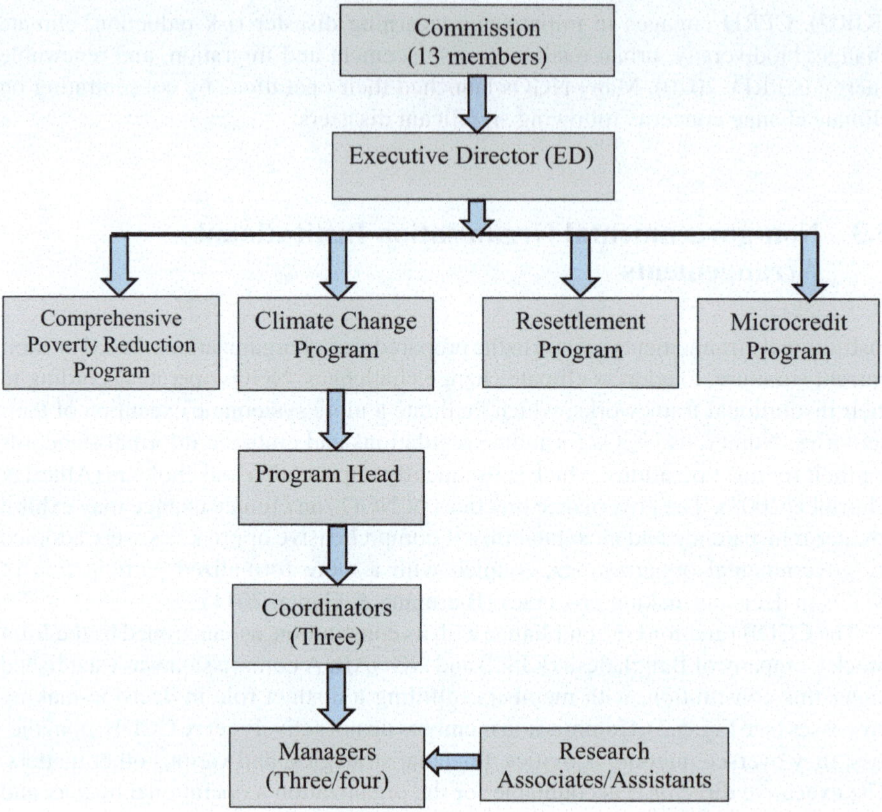

Fig. 5.2 Institutional framework of the Christian Commission for Development in Bangladesh. *Source* Hossain and Momtaz (2024)

numerous donor-funded projects and initiatives. The entities comprise the senior management team, an audit committee, a compliance committee, a gender forum, and independent procurement reviewers.

Upon receiving a project, CNRS requests approval from NGOAB to advance, detailing all facets of the research, including financial sources and budgetary considerations. The approval process of NGOAB is intricate and protracted; thus, CNRS secures money from INGOs or other sources prior to obtaining NGOAB clearance. CNRS must then await the implementation of a project. If the project receives support from the Fisheries Department or INGOs, project coordination and administration are conducted through the CNRS head office. This necessitates collaboration with the UNO at the sub-district level and with the district commissioner at the district level. CNRS participates in regular district meetings to delineate project advancements and address operational issues. CNRS has established protocols for monitoring and reporting difficulties at the field level. The executive board and general board meetings address the annual budget, program progress, and policy-level issues.

COAST Trust possesses a seven-member executive committee responsible for policy formulation, convening at a minimum of once every three months. It provides regular reports at both local and national government levels and conducts a yearly systematic review process in collaboration with internal and external stakeholders, including beneficiaries. The organization's accounts and financial systems undergo annual audits conducted by an external firm appointed by the NGOAB.

The DF possesses a 15-member executive committee, which holds the ultimate power for formulating action plans and policies (DF, 2020). A member of the executive committee serves as the secretary of the forum. Members of the executive committee frequently oversee the work of the forum's secretariat. A contingent of volunteers collaborates with core personnel to execute DF activities.

The governance framework of DUS fosters transparency and accountability throughout diverse projects and activities. The executive committee, elected by the general assembly, comprises professionals, activists, and entrepreneurs who utilize their skills and experiences to enhance DUS governance. The governing board, assisted by an audit committee, evaluates the organisation's financial status, the efficacy of its performance, and the reports from various sections to formulate suitable recommendations for action.

The surveyed NGOs adhere to robust institutional frameworks for their operations. These frameworks facilitate the functions of various entities and their specific responsibilities. Within these frameworks, these entities sustain relationships within the respective organization.

5.4 Non-governmental Organisation Climate Change Governance Practices

Governance of climate change has emerged as a significant concern for NGOs, which implement initiatives to tackle climate-related challenges. Non-governmental organizations face constraints in upholding accountability and transparency due to their significant reliance on donor funding; yet, governmental organizations allocate a considerable amount of resources to them (Allard & Martinez, 2008). Furthermore, diverse governance criteria, including accountability, transparency, and funding, grow intricate as NGOs extend their operations from local to global spheres (Young & Dhada, 2013). Lonnqvist et al. (as cited in Lopa & Ahmad, 2016) observed that 'the emphasis of Bangladeshi NGOs on good governance has become a priority for funders' (p. 783). NGOs collaborate extensively with local government agencies at the community level to enhance their climate change governance practices.

5.4.1 Non-governmental Organisation Accountability in Climate Change Activities

The Government of Bangladesh has a constitutional obligation to safeguard its citizens from natural calamities by bolstering community resilience. Numerous NGOs, alongside pertinent stakeholders, are concentrating on promoting enhancements, fostering constructive efforts, and augmenting governmental accountability (Bannerman et al., 2011). NGOs frequently have difficulties stemming from insufficient internal and external accountability and inadequate governance frameworks during their operations (Kim, 2003). Accountability has emerged as a significant concern for governmental and non-governmental organizations in Bangladesh (KII26). The accountability of NGOs in Bangladesh is associated with the engagement of several stakeholders, including beneficiaries, donors, and the government (Mir & Bala, 2014).

CPRD upholds accountability to both governmental entities and contributors. From its foundation, it prioritized climate change in conjunction with catastrophe management. CPRD developed a national plan addressing climate change. It additionally facilitated international discussions as a governmental delegation. There exists an alternative form of accountability regarding the utilization of funds. It allocates funds appropriately in compliance with project stipulations. CPRD generates two types of reports: a qualitative report detailing activity over a designated period and a financial statement outlining expenditures. The organization possesses multiple auditing systems. CPRD also compiles narrative and financial reports for the government to validate their fund disbursement. Consequently, it upholds accountability to all stakeholders, including development partners and governmental agencies. A responsible officer within a non-governmental organization elucidates the constraints associated with accountability:

> I am dissatisfied with the current method as it allows for the purchase of my accountability certificate. If I communicate with the relevant officials and allocate contribution funds for my projects, I can obtain the certificate with ease. The certificate of accountability is not the ideal embodiment of accountability. Accountability is contingent upon my deliverables and the extent of my acceptability by the stakeholders. If the pertinent stakeholders regard me as a competent researcher or my organization as a reputable research entity, then my accountability is commendable. Acquiring credentials related to accountability is fundamentally unjust. (KII08)

CCDB possesses a specific independent monitoring section and an official responsible for overseeing climate change-related initiatives. An officer is tasked with initiating and overseeing a project from its commencement. It has established monitoring objectives according to a specified structure (see to Table 5.4). The monitoring officer, located at the head office, is authorized to confidentially examine programs in various field-level offices. The monitoring officer is also required to oversee financial affairs. The monitoring system records observed progress over time concerning the project's metrics. The results are consistently communicated to contributors. CCDB is required to produce an annual progress report due to the intrinsic connection

Table 5.4 Christian Commission for Development in Bangladesh climate change project monitoring systems

Issues	Monitoring processes
Program implementation	Uses of different indicators based on the project proposal
	Submission of project reports to the senior management and donors
	Third-party auditing and evaluation undertaken after three years of the project's completion
	Monitoring of projects regularly at the local level
	Organising an national-level annual program on key achievements, learning and failures
Financial transaction	A good set of policies on financial issues
	Presence of transparent financial transaction systems
	A well-structured financial department
	Publishing an annual report on financial issues

Source Hossain and Momtaz (2024)

between a project's funding and its anticipated outcomes. A separate team within the NGOAB headquarters conducts internal audits to assess the efficient implementation of initiatives.

CCDB employs software known as Meaningful Interaction Analysis (MIA), which documents all data pertaining to projects and activities. Access to certain financial matters is limited under MIA, while aggregate financial data remains available. All CCDB projects will be monitored and managed using MIA software starting next year. A responsible official from an NGO articulates the accountability concerns pertaining to NGOs as follows:

> In my opinion, the accountability of climate projects is contingent upon the sorts of projects and the systems of project management employed. For instance, when we undertake a project with the Fisheries Department, our organization is responsible for ensuring accountability for that specific project. Effective project management and accountability systems are essential for ensuring that a project can advance with due diligence. Conversely, if accountability is not upheld and NGOs are chosen based on personal preferences and other criteria, then accountability cannot be assured. (KII09)

NGOs are accountable via their executive and board members. Depending on the size of the NGO, they may also uphold a hierarchical structure among their personnel inside their operational processes. For instance, CNRS has integrated accountability mechanisms from field to district-level officials. This encompasses the individual accountable at the headquarters or a project manager on-site. CNRS strives to uphold this accountability system across all its operations. Accountability is not just dependent on CNRS's systems; it is also associated with pertinent partners such as INGOs or GOs. International Non-Governmental Organizations (INGOs) exhibit greater rigor in upholding accountability and meticulously outline all aspects prior to project initiation. They also employ numerous indicators, including the scope of work, intended beneficiaries, and field-level execution procedures to ensure compliance.

Similar to other NGOs, the NCCB addresses climate change concerns and executes initiatives centered on Climate Change Adaptation (CCA). It possesses a distinct vision to enhance governance—particularly regarding climate change—by empowering the most disadvantaged populations. NCCB is striving to safeguard the rights and involvement of Bangladesh's climate-vulnerable populations while advocating for transparency in decision-making processes. A representative from a surveyed NGO observes:

> Many believe that NGOs lack accountability in their operations. Numerous organizations operate in Bangladesh; nevertheless, their accountability is not clearly defined. Our activities are overseen by four or five government agencies; nevertheless, how many departments are supervising the activities of others? Clarity must be upheld throughout all organizations to ensure accountability and achieve effective outcomes. (KII10)

NCCB endeavors to shape national climate policy to attain climate justice, ensuring that adaptation measures cater to the requirements of the most vulnerable community groups. It serves as a robust advocacy platform for its members and sustains connections with other civil society organizations (e.g. CANSA,[1] GOBESHONA[2]) and non-state actors (e.g. BAPA) to educate policy makers and promote climate justice and effective governance. It commenced a series of campaigns and lobbying efforts in the formation processes of the NAP to guarantee participation and the development of bottom-up, ecosystem-based local adaptation plans. It serves a crucial function through its lobbying, advocacy, and campaign efforts aimed at augmenting financial resources directed toward the most disadvantaged regions. It also underscores the transparent and efficient utilization of climate funding.

COAST Trust has an active, elected board of directors. Board meetings occur routinely, and all proposed projects undergo comprehensive evaluation before obligatory board approval. COAST Trust has encountered challenges in sustaining grassroots contact, particularly during the project design stages. It emphasizes the comprehensive advantages of a proposed initiative, rather than the benefits for specific groups.

Occasionally, while addressing climate change governance or policy matters, there was limited opportunity to garner substantial input from grassroots sources. In these instances, COAST Trust intensifies its focus on accountability during the screening and execution of projects and programs.

DUS, a local NGO in Hatiya, focuses on disaster education and socioeconomic development for marginalized coastal communities (DUS, 2020). This NGO engages in connection with the Upazila Disaster Management Committee, Union Disaster Management Committees, and various local NGOs.

---

[1] Climate Action Network South Asia (CANSA) is an association of over 286 civil society organizations working in South Asian countries to enhance government and individual effort to limit climate change-related effects.

[2] GOBESHONA is knowledge sharing platform for climate change-related issues in Bangladesh.

DUS functions in accordance with its organizational framework, requiring officials to report project progress to their respective superiors on a monthly, quarterly, semi-annual, or annual basis. Every DUS project encompasses qualitative and quantitative metrics and objectives. The DUS organizational structure comprises a monitoring cell and an audit cell to supervise and regulate the organization's actions. It possesses an annual strategy and budget, overseeing and evaluating its initiatives quarterly and annually. During its quarterly and annual evaluations, DUS discovers discrepancies from the established metrics and objectives of each approved project and analyzes the underlying causes of these variances. The executive committee typically evaluates the progress of their operations quarterly. They are then positioned to propose pertinent proposals or financing recommendations at the board meeting. DUS is divided into geographical zones and areas, with all pertinent personnel responsible for executing their designated projects in alignment with the sanctioned work program.

NGOs established the Association of Development Agencies (ADAB) to ensure responsibility among its members in Bangladesh (Kim, 2003). The assessed NGOs possess distinct accountability mechanisms, which their executive bodies utilize to oversee the organization. However, projects are occasionally overseen by NGOAB. Numerous NGOs fail to uphold effective accountability across all levels. Most NGOs, with the exception of larger organizations, typically do not adhere to climate finance governance strategies and generally lack accountability in the management of climate funding. Nevertheless, several NGOs presently prioritize responsibility in achieving their objectives.

5.4.2 Non-governmental Organisation Transparency in Climate Change Activities

NGOs are regarded as significant participants in international climate negotiations (Liu et al., 2017). Transparency is increasingly a critical requirement for both public and private climate governance frameworks, significantly influencing accountability and strengthening environmental outcomes (Baroi & Panday, 2018; Gupta & Mason, 2016; Hahn et al., 2015). It has been noted that NGOs engage in various concealed irregularities to secure funding from BCCT (see Table 5.5). TIB (2013a, 2013b) indicated that board members and executives of numerous NGOs are actively engaged in local politics, leveraging their political influence to obtain unwarranted advantages from government organizations. However, state legislation prohibits NGOs from engaging in partisan political activity (Lopa & Ahmad, 2016). NGOs lacking prior experience or administrative capacity for implementing climate change-related projects are excluded from consideration for climate grants. A local Bengali daily, The Kaler Kantho (27 May 2013), reported that numerous NGOs have acquired funding through political influence or by contravening established regulations (Bhuiyan, 2015). This excerpt addresses the transparency of NGOs.

Table 5.5 Selected non-governmental organisation irregularities from the Bangladesh Climate Change Trust Fund

Type of irregularities	Number of non-governmental organisations (NGOs)
NGOs that could not be traced	10
Use of current residence of chairperson/chief executive as a liaison office	3
Project obtained through political influence	9
Executive director or member of executive board involved in party politics	13
Embezzlement of funds from other projects	2
No office in the project area despite legal requirement	4
Inadequate number of meetings of NGO executive committee	8
Registration cancelled by the microcredit authority in the past	1

Source TIB (2013a, 2013b) and Bhuiyan (2015)

> Many NGOs encounter governance issues, particularly with cash utilization. Numerous NGOs were chosen to execute government-funded projects owing to political pressure in the past; nevertheless, their presence at the field level was not observed. The selection processes of NGOs lacked empirical evidence, and their capability to execute diverse projects was significantly deficient. Regrettably, other cases of misallocation were revealed over time by journalists and others. However, little actions were implemented by the relevant authority. (KII21) (KII26)

CPRD endeavors to uphold transparency and collaborates with government agencies on its research, advocacy, and recommendations. It often adheres to specific protocols during the execution of any research initiative and engages the community and pertinent stakeholders in dialogue. It also distributes reports to national-level stakeholders, including government organizations, NGOs, and civil society organization leaders, while disseminating its research findings. It is transparent in another regard: it disseminates reports to worldwide stakeholders by engaging in international conferences and seminars, utilizing these platforms to communicate its research findings. It guarantees project audits, disseminates audit reports on its website, and submits project reports to donors and governmental organizations.

This study examined two aspects related to CCDB: program implementation and financial review. CCDB is a reputable and competent organization. It upholds transparency in its financial dealings and has incorporated audit systems along with additional operational policies. It issues an annual financial report, fully transparent to the public. It adheres to governmental norms for non-governmental organization operations and reporting. The finance department is well-organized; nonetheless, financial statements remain nondigitized due to resource constraints. The CCDB financial systems are interconnected with banks and other financial entities. Consequently, this NGO upholds transparency in its climate change-related endeavors.

CCDB employs many management technologies to guarantee the effective execution of its activities. The CCDB internal project review mechanism is activated directly from a project's proposal to its completion. Upon project completion, a program monitoring team evaluates the project outcomes utilizing the indicators outlined in the initial project proposal. CCDB annually delivers project progress reports and conclusions to senior management and donors, based on the established indicators and objectives of the activity. CCDB performs third-party audits and evaluations, in conjunction with internal reviews, to comply with its standards and the stipulations of donors, when necessary. Within three years of a project's conclusion, CCDB has each project evaluated by a trained assessor. These evaluations are performed by accountable individuals at the operational level. CCDB collects reports, including comments and photographs from field officials, and consolidates them at the head office for dissemination to senior management and donors. It also disseminates project outcomes publicly, particularly to stakeholders and project participants. CCDB further hosts an annual national forum that emphasizes significant accomplishments and insights, along with challenges faced by the organization during the year.

CNRS employs a central monitoring mechanism termed a 'monthly learning sharing meeting,' wherein the designated project staff member delivers a quarterly work plan and a monthly progress report, emphasizing any associated problems. At the operational level, CNRS project personnel participate in monthly meetings to discuss challenges, strategies, accomplishments, and solutions. Personnel from this NGO frequently engage with local UNO and union officials to facilitate field operations. They also address issues stemming from these connections during monthly meetings with local administrative officials. Challenges encountered with local union representatives shall be addressed with the Upazila administration for resolution. CNRS additionally seeks authorization from local administrative officials to uphold CNRS rules for field operations.

The NCCB asserts that NGOs have a limited role in climate change matters, particularly in global talks and policy formulation. The government frequently seeks to avoid criticism from NGOs and CSOs over its climate change initiatives. NGOs collaborate with governmental entities or benefactors to execute diverse tasks. The absence of fundamental democratic values and practices is rendering the reliance on governmental support and collaboration problematic for both governmental organizations and non-governmental organizations. Certain NGOs have altered their operational tactics to avoid direct engagement with the government, instead functioning as subcontractors. These activities have constrained the capacity and efficacy of numerous NGOs in achieving beneficial objectives. Such NGOs possess limited opportunities for critical involvement or, alternatively, to endorse effective government efforts and propose constructive proposals for change and enhancement.

Since its inception in 2007, COAST Trust has consistently participated in COP, utilizing its own financial resources. A senior official of COAST Trust addresses the matter of transparency:

> We typically organize two news conferences prior to participating in the COP. During the pre-COP press conference, we underscore the necessity of addressing the expectations of vulnerable nations during the COP. We also incorporate alliance members and national members. We frequently organized photographic exhibitions focused on marginalized groups. We also participate in seminars regarding Bangladesh's adaptive capacities to access GCFs, adaptive technologies, loss, and damage, among other topics. We will summarize, in the post-COP press conference upon our return to Bangladesh, the expectations of various stakeholders, the accomplishments from the COP, and outline the pathways for future progress. (KII07)

COAST Trust has implemented protocols to ensure transparency. It requires backing from grassroots stakeholders, organizes conception meetings, and disseminates its initiatives. It also elucidates the rules for financial oversight during a project inception meeting with CSOs. It amplifies voices from various social sectors and engages a diverse array of individuals from local communities in inclusive problem-solving. A responsible officer from an NGO exemplifies the openness of climate money within NGOs as follows:

> We do not establish our own climate funding transparency and control procedures; instead, we enhance the capacity of civil society organizations to monitor each climate initiative effectively and efficiently. We consequently disclose all information and accounts pertaining to our budget, goals, programs, and the execution of numerous initiatives. We also disseminate the organization's audit report, financial statements, project ideas, and other relevant documents. We uphold a significant level of transparency in our programs and have garnered multiple accolades for our initiatives. (KII07)

DUS possesses a team responsible for conducting regular surveillance. A monitoring director delivers quarterly reports for each project. These papers address multiple obstacles to enable the pertinent authorities to resolve the concerns. DUS ensures transparency in its financial operations.

DUS possesses committees for social welfare and a microfinance regulatory body that evaluate program progress at bi-monthly meetings. Annual progress reviews are conducted, along with program deviations during the annual general meeting. Foreign benefactors and PKSF officials evaluate the advancement of DUS programs. These external audits also evaluate service performance and financial matters. DUS seems to possess robust transparency protocols. It disseminates program proposals and financial information to all stakeholders prior to commencing field operations. It meticulously focuses on project planning throughout monitoring and oversight tasks. DUS representatives concur that they do not anticipate exceptional outcomes from their programs.

The examined NGOs engage in several initiatives to promote openness. They now prioritize transparency more than they did previously. Nonetheless, their degrees of transparency vary due to their respective areas of coverage and the types of programs offered.

5.4.3 Stakeholder Participation in Non-governmental Organisation Activities

Climate change governance involves numerous social issues, so the significant involvement of pertinent stakeholders may contribute to addressing some of these challenges (Huitema et al., 2016; Kabiri, 2016). Ensuring the involvement of many stakeholders in NGO decision-making processes is crucial, while also acknowledging the pressures NGOs face from donor policies and international conventions (Lopa & Ahmad, 2016). Many government agencies tend to avoid collaboration with NGOs, despite the potential of certain large NGOs to substantially mitigate the impacts of climate change (Stock et al., 2020). The studied NGOs employed many methods to facilitate community and stakeholder engagement in their climate change initiatives.

CPRD guarantees the involvement of pertinent parties during the design process of its projects. It is presently executing a project about loss and damage from climate change in specific hotspots, recording the resultant effects, and engaging in discussions with the local community over the research. It is conveying its conclusions on loss and harm to national and global stakeholders. It disseminates project planning and results to other stakeholders. A mid-level employee of CPRD delineates the constraints on their involvement:

> We cannot guarantee the involvement of all stakeholders in climate policy research. We want to engage with the chosen stakeholders to examine climate policy and uncover policy deficiencies through more intuitive methodologies. (KII08)

Since 1972, CCDB has prioritized stakeholder involvement in its projects and programs. When formulating a project proposal, it identifies community-level stakeholders. It engages with these stakeholders to ascertain community requirements. It consistently disseminates the particulars of project plans. To do this, CCDB organizes several yard meetings and focus group discussions during the planning phase. It solicits opinions from foreign experts during the proposal drafting phase and disseminates project drafts to national-level specialists. The CCDB Lighthouse Project, for instance, concentrated on establishing entities like the Community Climate Resilience Centres (CCRC) (see Fig. 5.3). CCRC recognizes the participation of community groups in mitigating the impacts of climate change, leveraging local knowledge and resources, and strengthening community resilience strategies within its initiatives.

The NCCB asserts that the assessment of NGO participation extends beyond mere 'head counting' to encompass the involvement of suitable and proficient NGOs. Invited NGOs must possess pertinent experience to engage actively in climate change matters. This is contingent upon the nature of the NGO and their prior track record or expertise in this domain. This excerpt demonstrates the involvement of pertinent parties in climate initiatives conducted by NGOs:

> Numerous government initiatives necessitate consultation as a criterion for their execution. To fulfill consultation criteria, the successful NGO typically invites familiar participants, prioritizing attendance metrics. In a particular government initiative, 50 individuals engaged

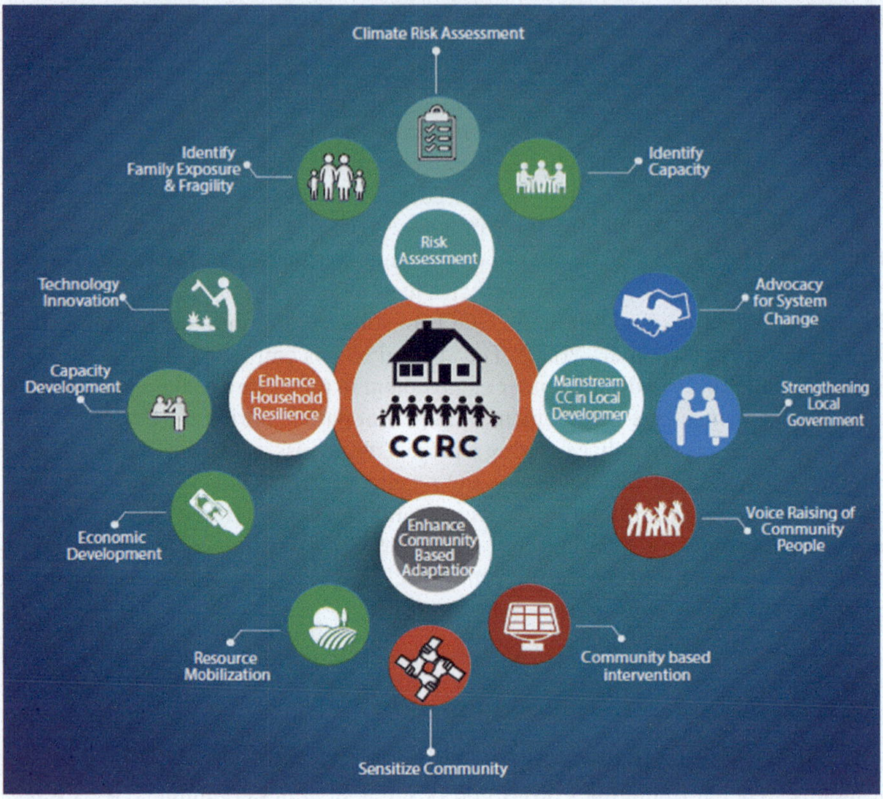

Fig. 5.3 Christian Commission for Development in Bangladesh project implementation framework. *Source* Christian Commission for Development in Bangladesh (2017)

> in the consultation processes and concurred that this did not constitute legitimacy. A suitable checklist would assist in evaluating the quality of prospective collaborating NGOs. (KII10)

Typically, CNRS adheres to distinct standards for each project. The programs primarily focus on natural resource management and climate change, benefiting those reliant on wetlands and forests. It identifies local stakeholders linked to certain projects. CNRS offers training in livelihood management, climate change adaptation, input assistance, and particularly in capacity building centered on the market system. It also identifies marginal farmers to engage in agricultural initiatives, including those farmers with small, excavated ponds capable of providing water year-round for crops and aquaculture. A senior CNRS official delineates this methodology:

> We possess the knowledge to identify project beneficiaries and to organize them into groups. We adhere to established protocols by identifying vulnerable individuals through communication with local government departments, categorizing them into underprivileged and severely needy groups based on the gathered information. (KII09)

COAST Trust aims to advocate and establish networks at both national and international levels to guarantee adequate participation. In 2008, it advocated for adequate involvement in BCCTF. It facilitates public engagement in many governmental and non-governmental initiatives. A mid-level staff member of a respected NGO articulates the following:

> Donors consistently exert pressure to guarantee public involvement in all projects. Upon examining the efficacy of help in a project, it becomes evident that public participation is essential. Many governmental and non-governmental organizations exhibit selective engagement in their project execution. We are addressing climate change governance issues with a mutually beneficial strategy through the Climate Justice Resilient Fund (CJRF[3]) and CFTM.[4] Consequently, we are regarded as a very people-oriented organization. (KII07)

DUS employs a bottom-up methodology for decision-making in initiatives concerning climate change and catastrophe management (see Fig. 5.4). Rural communities, by engaging in their programs, are incorporated as active participants in the decision-making process. All programs are disseminated to top local leaders. DUS solicits assistance from Union Disaster Management Committees regarding disaster management, climate change, or microfinance. DUS is a member of the Upazila Disaster Management Committee, with government departments including agriculture, fisheries, livestock, and engineering, such as the local government engineering department (LGED). It organizes round-table discussions to make decisions and execute policies and programs about development or disaster management. DUS is a member of the district-level DM committee. This allows DUS to promote bilateral communication between local communities and governmental groups for critical decision-making. DUS also cultivates networks with various organizations focused on climate change at both national and international levels.

DF asserts that community engagement is associated with local knowledge, and this association bolsters individuals' resilience in combating climate change. Participation is essential for remaining in impacted regions, acquiring comprehensive understanding of climate change challenges, and maximizing efforts to attain favorable results for vulnerable populations. Nonetheless, participation processes are not consistently executed accurately.

The surveyed NGOs are developing various strategies to facilitate stakeholder participation; nevertheless, not all are achieving adequate progress toward effective participation in their initiatives. Nevertheless, a minority are endeavoring to get stakeholder engagement. They obtain input from participants, which may disrupt program implementation processes. Consequently, they uphold minimal participation levels to circumvent procrastination.

---

[3] The Climate Justice Resilient Fund (CJRF) runs a donor-based project entitled 'Community-led Initiatives for Climate Justice and Resilience in the Islands and Coastal Areas of the Bay of Bengal in Bangladesh'; it was funded by COAST Trust, an NGO, in 2017. The main focus of this project was to enhance the lives and livelihoods of climate victims (CJRF, 2017).

[4] Climate Finance Transparency Mechanism is a project funded by the Department for International Development in Bangladesh in 2017. This project aims to increase the resilience of climate-vulnerable communities by enabling climate funds, and seeks to improve the level of transparency in governance (International Centre for Climate Change and Development, 2017).

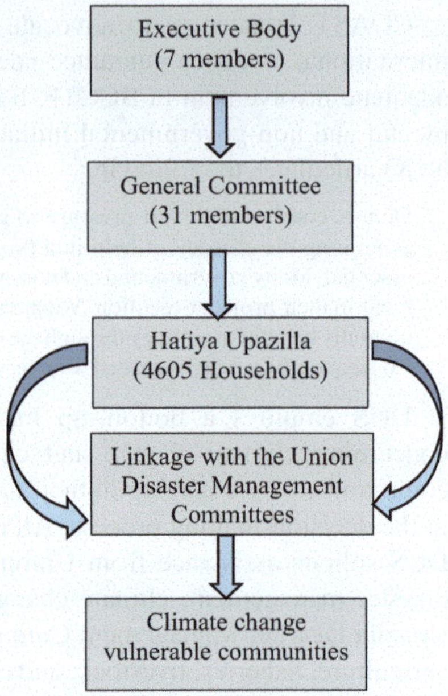

Fig. 5.4 Dwip Unnayan Songstha climate change-related decision-making procedures. *Source* Hossain and Momtaz (2024)

5.4.4 *Non-governmental Organisation Effectiveness in Climate Change Activities*

NGOs significantly optimize available resources and foster linkages between communities and governmental organizations (Hayes & Persha, 2010). Non-governmental organizations endeavor to implement innovations and enhancements for individuals in the most expedient manner. NGOs, often supported by donor funds and government initiatives, primarily address climate change concerns. Although there is a widespread absence of thorough evaluations to determine their efficacy, NGO activities evidently benefit individuals in numerous ways. Nonetheless, the documentation regarding the magnitude and efficacy of these advantages is inadequate. A lecturer at a public university elucidates the efficacy of NGO initiatives about climate change.

> Non-governmental organizations (NGOs) engage in various sectors, including agriculture, disaster relief, poverty alleviation, forestry, education, and health; however, many of them focus on microcredit, which implies a commercial side to their operations. The tasks of NGOs addressing climate change are regarded as business activities, encompassing research, implementation, and other development areas. Given that numerous NGOs prioritize their operations, we cannot anticipate improved results from them. (KII28)

Several NGOs, including CCDB and CNRS, are seeking partnerships with other organizations. They frequently conduct collaborative programs. These initiatives prevent program duplication, hence improving the efficacy of governance systems across all tiers. CCDB formulates its initiatives with regard to the priorities of local communities. It also evaluates the danger to prospective victims of climate change during the project's anticipated duration. It implements community governance frameworks to facilitate the efficient management of projects. Currently, CCDB is executing a project for a climate technology park.[5] This park showcases 50 technology addressing climate change, climate change adaptation, and mitigation. CCDB intends to progressively withdraw its sponsorship once this project can operate effectively without external assistance.

CPRD projects enhance government initiatives and strategic planning by addressing loss and damage attributable to climate change. CPRD also seeks to build national systems that address loss and damage in collaboration with the government. CPRD aims to encourage the government to establish definitive climate change protocols. CPRD has implemented an institutional framework to prevent redundancy among stakeholders. Initially, when the BCCTF was founded, NGOs anticipated that it would be overseen by an independent entity; regrettably, only two representatives from civil society organizations were included, while the majorities were from governmental departments (e.g., ministers, secretaries). Ministers are political representatives functioning inside political institutions, which led to the BCCTF not performing as anticipated.

NCCB advocates for a national climate strategy aimed at attaining climate justice, incorporating adaptation measures that cater to the needs of the most vulnerable populations, including the elderly, women, and children. Simultaneously, it fosters the transparent and efficient utilization of climate money. NCCB is the sole NGO dedicated exclusively to addressing climate change. Its objectives are distinctly defined, focusing on improved governance, particularly in relation to climate change. This empowers populations susceptible to climate change. It possesses attainable indicators or targets to direct project planning, hence facilitating the assessment of a project's success. Its aims encompass programs advocating for climate change.

COAST Trust has implemented many disaster preparedness and post-disaster rehabilitation initiatives in coastal regions. It focuses on financial sustainability, which is essential for the successful execution of initiatives. It circumvents donor-driven campaigns, as contributors may exhibit prejudice and alter COAST Trust's funding tactics.

When financing a climate initiative, it primarily targets vulnerable communities. The predominant portion of revenues generated by COAST Trust's microfinance initiatives is allocated to their campaigns. This is referred to as a 'rights-based approach.' Since 2007, it has effectively advocated for a national climate budget. It has actively engaged in numerous climate-related matters, including climate financing, embankment building, reform of the BWDB, fisheries, forestry, and climate-adaptive technologies.

---

[5] Climate Technology Park, a unique project of the Christian Commission for Development in Bangladesh, is focusing on climate change issue. This park is designed to present the five major ecological zones of Bangladesh—the coastal, dry, char land, hilly and haor (wet land ecosystem) zones.

DUS operates at the local level to execute plans and initiatives addressing climate change concerns. DUS possesses a contingency plan for disaster scenarios. In the event of a significant disaster, the established contingency plan is implemented. The lack of significant disasters post-2009 has diminished the efficacy and promptness of the initial strategy. DUS has recently expanded its initiatives into new domains not addressed by other NGOs, aiming to effectuate meaningful improvements in individuals' lives and livelihoods.

The efficacy of NGO initiatives with climate change issues fluctuates according to their varied programs. The assessed NGOs implement diverse programs according to established rules to guarantee successful results for users; nonetheless, the restricted financial availability has hindered program coverage.

5.4.5 Non-governmental Organisation Resource Dependency and Fund Mobilisation Strategies

Resources are an essential prerequisite for the existence and operational foundation of NGOs. Most NGOs function via donor-driven initiatives. Insufficient funding for NGOs frequently constrains both the quality and quantity of their projects (Viravaidya & Hayssen, 2001). Non-governmental organizations typically obtain financial resources for their regular operations from three primary sources (Mir & Bala, 2014) (see Fig. 5.5).

Securing sufficient money is consistently a formidable challenge for NGOs. NGOs are regularly contracted to execute government projects to secure operating financing, frequently collaborating with INGOs. International funding sources, both public and private, may be unable to sustain NGOs during global economic downturns, resulting in operational challenges for these organizations (Khieng, 2013). NGOs also obtain funding from governmental organizations to design and execute climate change-related initiatives. National-level NGOs hardly obtain direct funding. NGOs have obtained funding from PKSF, totaling USD 15.57 million by 2013, comprising USD 3.07 million from BCCTF and USD 12.5 million from BCCRF (TIB, 2013a, 2013b).

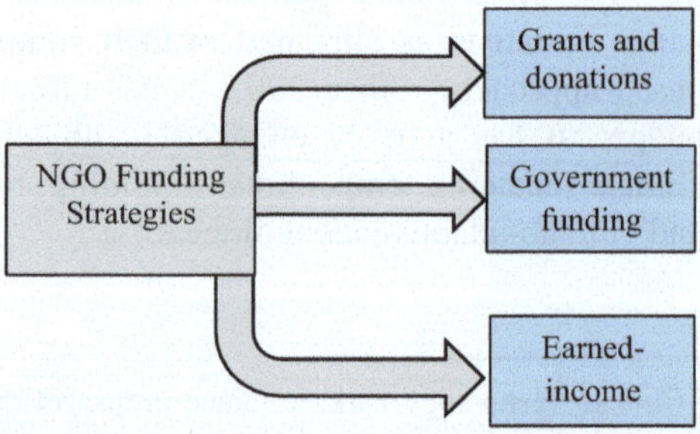

Fig. 5.5 Non-governmental organisation funding sources. *Source* Adopted from Froelich (1999) and Viravaidya and Hayssen (2009)

A primary issue associated with collaborating with INGOs is the substantial overhead expenses and considerable financial resources required to sustain offices and personnel. This results in less funding for grassroots initiatives. Competition among NGOs for national financing is intense, and available resources are frequently insufficient. NGOs obtain 10% of the total funds awarded by the Bangladesh Climate Change Trust Fund, which are distributed and managed by the Palli Karma-Sahayak Foundation (Rai et al., 2014). PKSF oversees community-level climate change initiatives for designated NGOs. In Dhaka-based NGOs, a significant percentage of their budget is allocated to head office management and personnel, resulting in insufficient funding for essential mission-related field projects. A responsible staff member of a NGO delineates the availability of funding for NGOs pertaining to climate change:

> Financial resources are not the issue for climate change-related initiatives; however, a well-structured managerial authority is essential to commence operations. When measures related to climate change become financially profitable, significant issues arise. The issue is often seen as significant; nonetheless, it lacks profitability and requires appropriate management. (KII10)

DUS possesses 40 years of expertise in securing financial assistance for initiatives addressing climate change concerns. It acquires funding from governmental entities like LGED, Education, Forest, and PKSF, as well as from international non-governmental organizations such as OXFAM, Canadian Aid, USAID, DANIDA, and JICA. DUS has obtained funding from the Forest Department (FD) for afforestation initiatives on a freshly constructed embankment along the coastal strip of Hatiya Island. COAST Trust, a local NGO, was addressing climate change issues and securing funding from donors, but it was required to navigate governmental organizations (e.g., NGOAB) to get those monies. A mid-level employee at a local NGO elucidates the fund disbursement protocols of non-governmental organizations.

> We obtained funding in accordance with the project's proposal. When the authorities announce funding opportunities, we typically submit our project proposal in response to the adverts. We obtain authorization from the relevant authority by adhering to the project's selection procedures. Subsequently, they disburse cash in accordance with policy, agreement, and duration, culminating in our receipt of the funds. (KII13)

Despite the increasing competitiveness of climate money for NGOs, they remain significantly reliant on donors, international non-governmental organizations (INGOs), and governmental organizations. Nonetheless, in addition to these problems, NGOs must rigorously adhere to accountability and transparency standards when utilizing donations from these sources, particularly multinational organizations (Table 5.6).

Table 5.6 NGOs' performance in climate change governance practices

Name of NGOs	Governance criteria				
	Accountability	Transparency	Participation	Government effectiveness	Resources
Centre for Participatory Research and Development	Determines the local requirements	Communicates feedback from grassroot to government level	Supports government bodies in developing plans and policies	Identifies the policy gaps	Encourages GOs to implement climate projects
Network on Climate Change, Bangladesh	Emphasizes the need for climate change advocacy programs	Works on climate change issues from local to global level	Ensures the participation of locals in climate change projects	Helps locals to enhance their resilience to climate change impacts	Focuses on best use of climate funds
Christian Commission for Development in Bangladesh	Works to build capacity of climate-vulnerable people	Contributes to global climate change negotiations	Arranges training programs for the young	Maintains an alliance to promote NGO activities	Monitors the financial matters properly
Centre for Natural Resources Studies	Improves the lives and livelihoods of climate-vulnerable people	Conducts activities from national to field level	Works on participatory climate vulnerability assessment	Introduces climate-smart agriculture	Focuses on fund mobilization strategies
COAST Trust	Campaigns for proper use of climate funds	Reinforces the capacity of CSOs	Develops adaptation strategies at the household and community level	Puts an inclusive voice on climate-integrated budget	Maintains transparency and effectiveness of budget

(continued)

Table 5.6 (continued)

Name of NGOs	Governance criteria				
	Accountability	Transparency	Participation	Government effectiveness	Resources
Disaster Forum	Ensures accountability of different agencies working for the people	Addresses the rights of different vulnerable groups	Makes collaboration with others to work on disaster management	Works as a catalyst to influence GOs and donors' polices	Ensures accountability in fund management
Dwip Unnayan Songstha (Island Development Association)	Conducts social afforestation program in 1990s	Operates a community radio called 'Radio Sagar Dwip' to warn coastal communities	Follows a contingency plan to manage disaster events	Works to promote saline-resilient crops	Receives and uses funds based on policy and agreement

Source Hossain and Momtaz (2024)

5.5 Non-governmental Organisation Successes in Climate Change-Related Activities

Edwards (as referenced in Frewer, 2013) observed that 'the proactive functions of NGOs significantly contribute to the growth of civil society, hence strengthening good governance' (p. 99). A dynamic civil society emerged in Bangladesh in the early 1970s. Over two decades, NGOs in Bangladesh have significantly influenced the global arena through comprehensive initiatives across multiple sectors (Chowdhury et al., 2013), including climate change (refer to Table 5.7). The NGO sector plays a crucial role in assisting climate-vulnerable populations in Bangladesh (Jenkins, 2003).

Numerous NGOs, like CPRD, operate on two tiers. Initially, NGOs assist the government directly in formulating plans and programs. Secondly, they provide grassroots feedback to government policymakers to enhance the government's response to local issues. NGOs do proactive research, including the identification of local policy deficiencies and necessities. Furthermore, they urge GOs to execute programs at the grassroots level.

COAST Trust has been involved in the oversight of BCCTF since its establishment and advocated for enhancements to BCCTF. It advocated for enhanced transparency, although achieved minimal success. It elicited apprehensions over the BCCTF's disbursement of funding, allocation, execution, and oversight. It has campaigned for a national climate-integrated budget since 2008. It facilitated the coordination of a comprehensive perspective on a climate-integrated budget. In response to this initiative, the government has developed several climate-integrated budget proposals in recent years.

The NCCB aims to amplify the voices of marginalized individuals about climate change matters at local, national, and international levels. Local communities possessed limited awareness of climate change in the past. Over the past 15 years, NCCB grassroots projects on climate change have equipped residents with fundamental knowledge and improved their capacity for action. The NCCB has formulated adaption techniques for both households and communities. The NCCB emphasizes that evaluating the performance of climate change projects necessitates an assessment of local engagement. It also underscores the necessity for climate change advocacy initiatives.

CNRS, an environmentally-oriented NGO, has effectively concentrated on disaster risk reduction, climate change adaptation, and mitigation via natural resource management initiatives. It illustrates several forms of climate-resilient agricultural systems and climate-smart agriculture, with assistance from the Food and Agriculture Organization (FAO) and local collaborators. CNRS has initiated several programs, like the Climate Resilient Ecosystems Livelihoods and Wetland Biodiversity Rehabilitation Project, which primarily emphasize climate change. It also created the Participatory Climate Vulnerability Assessment, which is executed at the village level. They have implemented micro-level initiatives. CNRS primarily concentrates on systems designed to enhance livelihoods. Its efforts encompass the demonstration

Table 5.7 Successes of surveyed non-governmental organisations in climate change-related activities

Name of NGO	Perceived successes
Centre for Participatory Research and Development	Identifies the policy gaps Supports government bodies in developing various plans and policies Determines the local requirements Communicates feedback from grassroot to government level
Network on Climate Change, Bangladesh	Works on climate change from local to global level Emphasises the need for climate change advocacy programs Helps locals to enhance their resilience to climate change impacts Works to ensure the participation of locals in climate change projects
Christian Commission for Development in Bangladesh	Works to build capacity of climate-vulnerable people Identifies the community needs based on science and statistics Developing a 'technology park' incorporating various measures related to adaptation and mitigation Arranges training programs for the young Contributes to global climate change negotiations Maintains an alliance to promote non-government organisation activities
Centre for Natural Resources Studies	Introduces climate-smart agriculture Operating various projects to improve the lives and livelihoods of climate-vulnerable people
COAST Trust	Campaigns for proper use of climate funds Helps coordinate an inclusive voice on a climate-integrated budget Develops adaptation strategies at the household and community level
Disaster Forum	Making collaboration with others to work on disaster management Works as a catalyst to influence GOs and donors' polices for making a conducive environment for NGOs Addresses the rights of different vulnerable groups Works to ensure accountability of different agencies working for the people

(continued)

Table 5.7 (continued)

Name of NGO	Perceived successes
Dwip Unnayan Songstha (Island Development Association)	Conducted social afforestation program in 1990s Follows a contingency plan to manage disaster events Works to promote saline-resilient crops Operates a community radio called 'Radio Sagar Dwip' to warn coastal communities before disasters

Source Hossain and Momtaz (2024)

of several climate-smart agricultural systems, including aquaculture, rapid-growth farming species, vertical gardening, sage cultivation, and crop diversification. The results of these programs are inadequately documented and have not been evaluated in terms of their impact on local livelihoods and the enhancement of CCA measures. CNRS engages in numerous supplementary initiatives at the field level.

CCDB aims to enhance the capabilities of those residing in marginalized communities and those operating at the national level. They assess community needs through scientific methods and statistical analysis at the community level. They do risk assessments to organize support and resources, prioritizing the most vulnerable during catastrophes. It is in the process of establishing a technological park, where it intends to implement 50 distinct technologies, including 20–30 associated with climate change adaptation and mitigation. CCDB offers instruction to numerous young individuals to augment their capabilities. It disseminates documents regarding climate change matters. CCDB additionally participates in international climate change negotiations. It has participated in conferences, including the COP, since 2014. It engages in advocacy initiatives and collaborates with NCCB, a consortium of NGOs focused on climate change. It functions for climate-resilient construction at both local and national tiers.

DUS operates on Hatiya Island and emphasizes Climate Change Adaptation (CCA). The inaugural social afforestation initiative in this region was commenced in the 1990s. DUS possesses a contingency plan for disasters, which is implemented during operations. They exhibited and advocated for saline-resistant crops and high-yield crop varieties (see Table 5.8). They also underscore the impact of climate change on the local fisheries, poultry, and cattle industries in initiatives aimed at enhancing awareness among the community.

DUS manages a community radio station named 'Radio Sagar Dwip', which engages island residents with concerns related to natural catastrophes by disseminating news and programs (see Fig. 5.6). The station provides feedback to communities susceptible to disasters regarding coping measures. A significant portion of the population in Hatiya consists of fisherfolk who depend on deep-sea fishing for their sustenance. The station provides the island's residents, as well as those from adjacent

Table 5.8 Dwip Unnayan Songstha activities relating to climate change adaptation

Issues	Number of beneficiaries
Raising the plinth above the flood level	800
Introducing saline-resilient crop species	3000
Managing cyclone shelters	20,000
Developing social afforestation programs	1500
Warning fishers during disasters	1,000,000
Improving lives and livelihoods through various programs	20,000

Source Hossain and Momtaz (2024)

Fig. 5.6 Radio Sagar Dwip's logo. *Source* http://dusbangladesh.org

islands, with information and programming about entertainment, disaster preparedness, agriculture, health, news, and local weather predictions. Radio Sagar Dwip plays a crucial role in assisting local authorities to adopt a proactive approach during emergencies. DUS instills optimism in the islanders through its many catastrophe preparedness initiatives.

DF, a nationwide disaster preparedness network centered in Dhaka, has collaborated with 70 other organizations and individual activists on sustainable disaster management. It aims to amalgamate disaster management with other developmental initiatives (DF, 2020). It functions as a catalyst to shape government and donor policies, fostering a conducive atmosphere for NGOs. It pertains to the rights of various disaster-prone groups and advocates for alternative viewpoints on disaster mitigation. Furthermore, it strives to guarantee the responsibility of many entities operating inside and for local communities.

5.6 Non-governmental Organisation Challenges in Climate Change-Related Activities

The studied NGOs have numerous hurdles in addressing climate change issues (see Table 5.9). Securing funding poses a significant difficulty for NGOs, as obtaining financial resources through project proposals is rarely a straightforward endeavor. Five of the seven surveyed NGOs expressed their apprehensions regarding funding.

Initially, an NGO must engage with prospective partners to secure project funding and persuade donors that the planned initiative corresponds with the partners' interests. Barman (as referenced in Mir & Bala, 2014) identified a comparable issue, observing that "a primary challenge for NGOs is the misalignment between their objectives and the intentions of their funders" (p. 1844). Secondly, NGOs must guarantee timely delivery. When an NGO receives funding from a donor, it must secure appropriate approvals from government organizations. This endeavor is consistently arduous, requiring substantial time, effort, and financial resources.

The Government of Bangladesh gives annual funds for Climate Change Adaptation (CCA) initiatives. A significant share of climate cash is designated for various government departments, while a lesser amount is distributed to NGOs by PKSF. Due to constrained funding for NGOs, PKFS often favors micro-level projects or investments. The enduring viability of NGO operations is a significant barrier for efforts related to climate change adaptation (Stock et al., 2020). Bhuiyan (2015) observed that numerous NGOs pursued funding for climate change-related initiatives, yet failed to satisfy the requisite requirements.

A distinct difficulty for NGOs is the lack of transparency in the recruitment processes for diverse projects. This may foster the notion (and actuality) that nepotism significantly influences the selection process, due to the personal connections of stakeholders and political pressure (Rai et al., 2014). Several sanctioned projects were terminated following media coverage. The subsequent excerpt elucidates this:

> Individuals are susceptible not only to climatic influences but also to a range of contextual elements, including policies, processes, and institutions. These are not clearly articulated in climate initiatives. Consequently, our accomplishments throughout the project are rudimentary and unsustainable due to a lack of awareness or failure to address contextual difficulties. (KII09)

Personnel are regarded as the fundamental resource of NGOs (Gauri & Galef, 2005). A significant number of staff members hired by NGOs lack substantial experience due to the inadequate remuneration provided. To attract high-quality personnel, NGOs must offer competitive compensation. INGO personnel possess superior skills and certifications, resulting in elevated compensation.

NGOs address climate change challenges under the overarching framework of 'civil society'; nevertheless, the availability of project funding under this expansive category is diminishing. A number of donors base their funding decisions on technical criteria such as cost-effectiveness and the delivery of specific objectives. If governmental organizations and donors restrict NGOs to service delivery initiatives and exclude them from participating in or supporting policy formation, the capacity of NGOs to effectuate meaningful reforms would be significantly impaired.

The participation of NGOs in the formulation of national plans or policies remains a significant dilemma. In the absence of NGO participation, such publications transform into governmental proclamations that may not accurately represent the national agenda. NGOs lack ownership of government policy documents, as these rarely contain or reflect NGO proposals, observations, requests, and recommendations. As previously said, NGOs typically operate at the community level, necessitating the

Table 5.9 Challenges of surveyed non-governmental organisations in climate change-related activities

Name of non-governmental organisation (NGO)	Perceived challenges
Centre for Participatory Research and Development	Lack of adequate funding for various programs Less recognition as an NGO than government organisations receive Maintaining accountability and transparency
Network on Climate Change, Bangladesh	Generally little scope to be involved in policy-level matters NGOs' roles are limited in service delivery Limited roles in climate change issues especially in international negotiations, policy-making and in other cases Participation is always a significant issue Lack of sufficient funds
Christian Commission for Development in Bangladesh	Lack of proper involvement in community capacity building Lack of adequate funding from the donors and government departments
Centre for Natural Resources Studies	Lack of adequate funding NGO selection processes for funding is not transparent Many projects are of very short duration Lack of required data in terms of production and adaptation efficiency Few skilled staff
COAST Trust	Limited spaces for campaigns To maintain transparency and effectiveness of budget Lack of government support in many aspects
Disaster Forum	Duplicated programs focusing on climate change issues Lack of investment for the required research and innovation Lack of proper participation of the community people
Dwip Unnayan Songstha (Island Development Association)	Increased number of disaster events occurring at the local level Lack of effective participation of NGOs in policy formulation Lack of adequate support from government officials Lack of political commitment Lack of public awareness

Source Hossain and Momtaz (2024)

involvement of local stakeholders. In the absence of substantial public participation in the execution of climate change-related initiatives, the likelihood of achieving meaningful results is minimal.

Despite the surveyed NGOs prioritizing accountability and transparency in their initiatives, they frequently struggled to uphold these principles during field-level execution. They are dedicated to upholding honesty and accountability in the utilization of donor cash. The government possesses rules to oversee the openness and accountability of grant utilization; nevertheless, in numerous instances, these governance practices are not adhered to in NGO initiatives.

A primary difficulty for NGOs is sufficient capacity. Non-governmental organizations primarily concentrate on short-term initiatives that entail local interventions, rather than on long-term sustainable projects. Infrastructure development is typically impractical for NGOs. The Cyclone Preparedness Programme (CPP) functioned effectively with disaster early warning systems; nonetheless, it necessitated interaction with governmental operations. The viability of NGO operations concerning CCA-related activities is evidently problematic. This excerpt illustrates the problems associated with these challenges:

> The majority of NGOs experience significant conflicts of interest, leading them to prioritize their microfinance beneficiaries when selecting recipients, regardless of whether these individuals belong to climate-vulnerable groups. This is very regrettable for the at-risk neighborhoods. The entire community anticipates the advantages of a specific initiative, as it has been sanctioned for communal benefit. Non-governmental organizations tend to serve a certain demographic rather than addressing the needs of the entire population. The supervision methods designed to mitigate conflicts of interest and enhance the project's success for the entire community generate controversy over the actions of NGOs. (KII26)

The absence of databases documenting production efficiency and adaptation efficiency is a significant concern. Upon the occurrence of a disaster, numerous management activities are implemented. The subsequent excerpts delineate the obstacles faced by NGOs concerning climate change initiatives in Bangladesh.

> Numerous initiatives involving NGOs are classified as climate-related projects. Our constraints often prevent us from recognizing our necessary actions. We frequently attempt to communicate our worries to donors and should endeavor to modify our behaviors and activities to better line with their objectives. We successfully finished a 'crop diversification' project, which we are now designating as a 'salient-resilient crops' project. Diverse initiatives are occurring, although the critical analysis of these topics is not being conducted adequately. We are unable to discern our requirements in this context. We are often unaware of our own limitations. (KII10)

> We lack sufficient databases on the climate variations of a certain region. Individuals are not receiving the appropriate individuals and suitable packages. Reports indicate that following Cyclone Aila,[6] afflicted individuals were required to pay for benefits in certain locations. We lack a comprehensive database of climate change-vulnerable regions and populations, together with their specific locations. This database must be updated regularly. Requesting a list of vulnerable individuals from a Union Council office may yield an inaccurate compilation. If we have an optimal list, we can make informed decisions and assist the actual victims promptly during and after crisis occurrences. (KII09)

[6] Cyclone Aila hit the coastal part of Bangladesh on 25 May 2009. It affected 3.9 million people and caused the death of 190 people (Asaduzzaman, 2015).

A significant difficulty for NGOs is the lack of national-level platforms for campaigns aimed at promoting climate change initiatives. Occasionally, NGOs are compelled to adhere to governmental directives, resulting in alterations to their campaign methods.

Any NGO that gets government funding for climate change-related initiatives undergoes project approval without evaluating the impact assessment of the specific project. Occasionally, initiatives are assessed superficially. Officials from the questioned NGOs indicated restricted access and assistance from BCCTF and BCCRF, which are respectively backed by development partners and the Government of Bangladesh. A climate change specialist delineates the functions of non-governmental organizations:

> The subject of climate change has not yet received significant attention within the NGO sector. Several NGOs are engaged in renewable energy initiatives, enhancing energy efficiency in factories, and addressing other related concerns. Nonetheless, these are categorized on the mitigation side. Flood control in the context of adaptation is exclusively a public-sector endeavor. Consequently, NGOs can forward additional projects and initiatives addressing climate change concerns. (KII22)

In certain instances, NGOs can uphold governance standards during project execution. Nonetheless, public understanding of climate change governance is a considerable difficulty at the community level. Local populations frequently lack the requisite information to comprehend the intricacies of climate change. They rely significantly on local administration and NGOs for access to services from climate change-related initiatives. Consequently, companies may be unable to capitalize on the prospective advantages of climate change initiatives.

5.7 Discussion and Conclusion

Bangladesh is in the forefront of global climate change-induced disasters. Historically, NGOs have focused on environmental or climate change issues and collaborate closely with government organizations. Although most governmental organizations (GOs) exhibit a lack of interest in collaborating with NGOs, the latter favor partnership. Numerous NGOs are diligently striving to mitigate climate change-induced vulnerabilities for the populace of Bangladesh (Chowdhury et al., 2013; Stock et al., 2020). They implement initiatives that bolster grassroots efforts in the nation's climate-sensitive regions. These enhance individuals' lives and livelihoods, significantly fortifying local organizations (Young & Dhada, 2013). NGOs are enhancing the resilience of local populations in disaster management, thereby mitigating the impacts of climate change at both national and local scales.

NGOs and civil society organizations (CSOs) are integral to governance systems and serve as a catalyst for international negotiations through proactive mobilization. NGOs in Bangladesh also fulfill significant responsibilities in Conferences of the Parties (COPs). Prior to participating in COPs, NGOs deliberated on various facets of the negotiating processes and assisted the government in its pursuit of achieving successful results from the COPs. NGOs emphasized that the government must uphold honesty during negotiation processes. Moreover, NGOs have facilitated community sensitization and knowledge on climate change issues. Consequently, Bangladesh has achieved notable results at the governmental level, facilitating the effectiveness of community-based adaptation.

A multitude of NGOs operate in diverse capacities in Bangladesh, with a growing emphasis on profit generation. Edwards (1999) contends that the NGOs in Bangladesh prioritize their own growth and development over delivering assistance to the impoverished. Many perceive their actions as commercial rather than as service-oriented. Numerous NGOs have failed to achieve the anticipated standard of governance in their operations (Kabeer et al., 2012); yet, the surveyed NGOs are implementing governance in their climate change-related initiatives. NGOs have exhibited a diminished propensity to spend in regions with minimal prospects for profits and in initiatives that are unlikely to provide services at the anticipated standard. Moreover, NGOs encounter challenges in sustaining effective climate change governance practices at the grassroots level due to the protracted processes necessary for realization. Numerous long-term projects fail to yield the expected outcomes as intended.

Historically, monitoring instruments concerning governance issues in NGO operations have been underutilized; nevertheless, several national NGOs are now employing sophisticated monitoring techniques. These are infrequently utilized in NGO operations. In numerous instances, they are excluded from governmental initiatives. NGOs frequently disseminate these tools to governmental organizations. Nonetheless, significant issues persist in decision-making systems, including the allocation of funding and the beneficiaries of climate change initiatives, as well as the integration of these governance mechanisms into mainstream practices.

NGOs encounter difficulties in adhering to accountability protocols in climate change-related initiatives. NGOs uphold accountability to both governmental entities and funders, while also being subject to oversight by their internal audit teams. NGOs underscore accountability to substantiate their operations across many levels. Transparency has become an essential concern for NGOs in the management of climate change-related initiatives. Despite numerous inconsistencies exposed by the media, many beneficiaries of NGO services have expressed gratitude for their endeavors.

NGO efforts are essential to address the issues posed by climate change, in conjunction with the involvement of other stakeholders. In the absence of these actions, the impacts of climate change will undermine the substantial resources and efforts being allocated. NGOs operating at the community level must prioritize the involvement of pertinent local stakeholders. NGOs frequently encounter difficulties in securing necessary funding from governmental organizations and benefactors. They must contend with others, and this rivalry has intensified. NGOs have endeavored to shape the allocation of cash for their operations. NGOs functioning in rural regions are improbable to allocate greater funds for office administration compared to those operating in urban settings. Consequently, companies adjust their project budgets, restricting financial resources for at-risk locations and populations.

The surveyed NGOs function according to their institutional frameworks, which differ based on the size of the NGO and its geographical scope. Typically, NGOs are administered by either an executive committee or an autonomous entity. These entities are tasked with supervising operational procedures, organizational efficacy, and project outcomes. This oversight frequently improves NGO operations, guaranteeing substantial results.

The efficient climate governance methods of NGOs are crucial for securing the active involvement of pertinent stakeholders. This promotes effective and sustainable growth processes. While it is improbable that NGOs will engage in development activities analogous to GOs, they are making substantial efforts to bolster local communities' capacity to combat climate change, thereby safeguarding lives and livelihoods from the detrimental impacts of climate change-induced disasters. The vigorous efforts by NGOs at the grassroots level emphasize climate change adaptation and mitigation. This will result in substantial transformations within this sector.

References

Ahmad, M., & Rahman, A. (2011). *Climate of coastal cooperation: Stimulating role of NGOs in Bangladesh.*

Allard, G., & Martinez, C. A. (2008). *The influence of government policy and NGOs on capturing private investment*. Paper presented at the OECD Global Forum on International Investment VII, Paris, France.

Asaduzzaman, M. (2015). *Livelihood vulnerability of women in the context of climate change impacts: Insights from coastal Bangladesh* [Unpublished doctoral thesis], University of Newcastle, Australia.

Bannerman, M., Rashid, M. H. O., & Rejve, K. (2011). *NGO-government partnerships for disaster preparedness in Bangladesh.* https://odihpn.org/magazine/ngo%c2%96government-partnerships-for-disaster-preparedness-in-bangladesh/

Baroi, H. S., & Panday, P. K. (2018). NGOs in local governance in Bangladesh. *South Asian Survey, 22*(2), 156–170. https://doi.org/10.1177/0971523117744149

Bhuiyan, S. (2015). Adapting to climate change in Bangladesh. *South Asia Research, 35*(3), 349–367. https://doi.org/10.1177/0262728015598702

Biermann, F., & Gupta, A. (2011). Accountability and legitimacy in earth system governance: A research framework. *Ecological Economics, 70*(11), 1856–1864. https://doi.org/10.1016/j.ecolecon.2011.04.008

CCDB. (2017). *Being resilient: Addressing climate change evidences on the outcome and learning of Bangladesh lighthouse project.* https://ccdbbd.org/

Centre for Natural Resources Studies. (2020). Centre for Natural Resources Studies. https://www.cnrs.org.bd/cnrs.html

Centre for Participatory Research and Development. (2020). Centre for Participatory Research and Development. https://cprdbd.org/

Chowdhury, A. M. R., Bhuiya, A., Chowdhury, M. E., Rasheed, S., Hussain, Z., & Chen, L. C. (2013). The Bangladesh paradox: Exceptional health achievement despite economic poverty. *Lancet, 382*(9906), 1734–1745. https://doi.org/10.1016/s0140-6736(13)62148-0

CJRF. (2017). *Community-led initiatives for climate justice in Bangladesh.* https://www.cjrfund.org/coast

Disaster Forum. (2020). *Foundation for disaster forum: A national network on disaster preparedness and risk reduction.* https://www.disasterforum.org/index.html

Dwip Unnayan Songstha. (2020). *Dwip Unnayan Songstha*. http://dusbangladesh.org/get-involved/
Edwards, M. (1999). NGO performance—What breeds success? New evidence from south Asia. *World Development, 27*(2), 361–374.
Frewer, T. (2013). Doing NGO work: The politics of being 'civil society' and promoting 'good governance' in Cambodia. *Australian Geographer, 44*(1), 97–114. https://doi.org/10.1080/00049182.2013.765350
Gauri, V., & Galef, J. (2005). NGOs in Bangladesh: Activities, resources, and governance. *World Development, 33*(12), 2045–2065. https://doi.org/10.1016/j.worlddev.2005.05.009
Gupta, A., & Mason, M. (2016). Disclosing or obscuring? The politics of transparency in global climate governance. *Current Opinion in Environmental Sustainability, 18*, 82–90. https://doi.org/10.1016/j.cosust.2015.11.004
Hahn, R., Reimsbach, D., & Schiemann, F. (2015). Organizations, climate change, and transparency. *Organization & Environment, 28*(1), 80–102. https://doi.org/10.1177/1086026615575542
Hayes, T., & Persha, L. (2010). Nesting local forestry initiatives: Revisiting community forest management in a REDD+ world. *Forest Policy and Economics, 12*(8), 545–553. https://doi.org/10.1016/j.forpol.2010.07.003
Hossain, M. F., & Momtaz, S. (2024). Climate Change Governance Practices of Non-governmental Organisations in Bangladesh. *The Dhaka University Journal of Earth and Environmental Sciences, 13*(1), 105–119. https://doi.org/10.3329/dujees.v13i1.77580
Huda, K. A. S. M. N. (2020). Micro-credit, trust, and social solidarity in Bangladesh: A socio-philosophical analysis. *Cultural Dynamics, 32*(4), 282–306. https://doi.org/10.1177/0921374020920152
Huitema, D., Adger, W. N., Berkhout, F., Massey, E., Mazmanian, D., Munaretto, S., Plummer, R., & Termeer, C. J. A. M. (2016). The governance of adaptation: Choices, reasons, and effects. Introduction to the special feature. *Ecology and Society, 21*(3), Article 37. https://doi.org/10.5751/es-08797-210337
Ilyas, M. (2010). *Legal and regulatory framework of NGOs in Bangladesh: Critical overview and my cogitations*. https://iliyasmohammed.wordpress.com/2010/09/01/legal-and-regulatory-framework-of-ngos-in-bangladesh-critical-overview-and-my-cogitations/
Intergovernmental Panel on Climate Change. (2012). *Managing the risks of extreme events and disasters to advance climate change adaptation*. https://www.ipcc.ch/pdf/special-reports/srex/SREX_Full_Report.pdf
International Centre for Climate Change and Development. (2017). *Brief background paper on climate finance transparency mechanism project*. http://www.icccad.net/climate-finance-transparency-mechanism-cftm/
Jenkins, B. (2003). NGOs in Bangladesh: Are they in a strong position to assist vulnerable people living on low-lying lands to cope with floods? *Journal of South Asian Studies, 26*(3), 455–466. https://doi.org/10.1080/0085640032000178989
Kabeer, N., Mahmud, S., & Isaza Castro, J. G. (2012). NGOs and the political empowerment of poor people in rural Bangladesh: Cultivating the habits of democracy? *World Development, 40*(10), 2044–2062. https://doi.org/10.1016/j.worlddev.2012.05.011
Kabiri, N. (2016). Public participation, land use and climate change governance in Thailand. *Land Use Policy, 52*, 511–517. https://doi.org/10.1016/j.landusepol.2014.12.014
Khieng, S. (2013). Funding mobilization strategies of nongovernmental organizations in Cambodia. *VOLUNTAS: International Journal of Voluntary and Nonprofit Organizations, 25*(6), 1441–1464. https://doi.org/10.1007/s11266-013-9400-7
Kim, J. (2003). *Accountability, governance, and non-governmental organizations: A comparative study of twelve Asia-Pacific Nations*. Paper presented at the Governance, Organizational Effectiveness, and the Non-Profit Sector in Asia Pacific, Manila, Philippines. http://citeseerx.ist.psu.edu/viewdoc/download?doi=10.1.1.135.9520&rep=rep1&type=pdf
Liu, L., Wang, P., & Wu, T. (2017). The role of nongovernmental organizations in China's climate change governance. *Wiley Interdisciplinary Reviews: Climate Change, 8*(6), E483. https://doi.org/10.1002/wcc.483

Lopa, F. G. R., & Ahmad, M. M. (2016). Participation of CSOs/NGOs in Bangladeshi climate change policy formulation: Co-operation or co-optation? *Development in Practice, 26*(6), 781–793. https://doi.org/10.1080/09614524.2016.1200536

Mir, M., & Bala, S. K. (2014). NGO accountability in Bangladesh: Two contrasting cases. *VOLUNTAS: International Journal of Voluntary and Nonprofit Organizations, 26*(5), 1831–1851. https://doi.org/10.1007/s11266-014-9513-7

NGO Affairs Bureau. (2020). *List of all NGOs*. http://www.ngoab.gov.bd/site/page/3de95510-5309-4400-97f5-0a362fd0f4e6/

Program Development Office for Integrated Coastal Zone Development Plan (2005) *Investment and financing strategy for coastal zone development in Bangladesh*. http://warpo.portal.gov.bd/sites/default/files/files/warpo.portal.gov.bd/page/aa04373f_0ca3_49a5_b77e_5108186638dc/wp037.pdf

Rahman, S. (2016). Development, democracy and the NGO sector. *Journal of Developing Societies, 22*(4), 451–473. https://doi.org/10.1177/0169796x06072650

Rahman, M. M., & Huang, D. (2019). Climate governance initiatives: Snapshots from Bangladesh. *Journal of Geoscience and Environment Protection, 7*(9), 131–147. https://doi.org/10.4236/gep.2019.79010

Rai, N., Huq, S., & Huq, M. J. (2014). Climate resilient planning in Bangladesh: A review of progress and early experiences of moving from planning to implementation. *Development in Practice, 24*(4), 527–543. https://doi.org/10.1080/09614524.2014.908822

Raustiala, K. (2001). Nonstate actors in the global climate regime. In U. Luterbacher & D. F. Sprinz (Eds.), *International relations and global climate change* (pp. 95–116). MIT Press.

Stock, R., Vij, S., & Ishtiaque, A. (2020). Powering and puzzling: Climate change adaptation policies in Bangladesh and India. *Environment, Development and Sustainability, 23*, 2314–2336. https://doi.org/10.1007/s10668-020-00676-3

Thomalla, F., Cannon, T., Huq, S., Klein, R. J. T., & Schaerer, C. (2005). *Mainstreaming adaptation to climate change in coastal Bangladesh by building civil society alliances*. Paper presented at the Solutions to Coastal Disasters 2005, Charleston, USA.

Transparency International Bangladesh. (2013a). *Climate finance in Bangladesh: Governance challenges and way out (executive summary)*. https://www.ti-bangladesh.org/beta3/images/max_file/es_cfg_executive_summary_english.pdf

Transparency International Bangladesh. (2013b). *Climate finance in Bangladesh: Governance challenges and way out*. https://www.ti-bangladesh.org/beta3/index.php/en/activities/4086-tib-demands-effective-measures-to-ensure-governance-and-reduce-corruption-in-climate-finance-2

Viravaidya, M., & Hayssen, J. (2001). *Strategies to strengthen NGO capacity in resource mobilization through business activities*. https://www.unaids.org/sites/default/files/media_asset/jc579-strategies_ngo_en_2.pdf

Young, S. T., & Dhada, K. K. (2013). Role of governments and nongovernmental organizations. In S. T. Young & K. K. Dhada (Eds.), *Sustainability: Essentials for business* (pp. 214–242). Sage.

Zafar Ullah, A. N., Newell, J. N., Ahmed, J. U., Hyder, M. K., & Islam, A. (2006). Government-NGO collaboration: The case of tuberculosis control in Bangladesh. *Health Policy and Planning, 21*(2), 143–155. https://doi.org/10.1093/heapol/czj014

Chapter 6
Community Perceptions of Climate Change Governance

6.1 Introduction

This chapter presents local communities' perceptions of climate change governance practices, considering aspects such as accountability, transparency, participation, governance effectiveness and resources. In this chapter, the following research question will be answered: 'What is the level of participation of local communities in climate change governance practices'. This chapter also shows the demographic characteristics of participants in the selected communities, including their age structure, educational level, marital status, occupational patterns and housing types. A discussion follows about the respondents' perceptions of climate change, the relationships of stakeholders, and climate change governance practices at the local level (based on selected criteria). The chapter then examines challenges at the local level. For data collection purposes, a semi-structured questionnaire was used to interview household heads and KIs, and an open-ended FGD checklist was used in the study areas.

6.2 Community Age Structure

Age structure in a community is an important criterion to understand its human development potential. The basic criterion for selecting the heads of households was that they be adults of working age (18–65 years old). In addition, they needed to perceive climate change issues in their areas. Most respondents were born and raised in the selected study sites. The head-of-household ages were divided into six categories: 18–24, 25–34, 35–44, 45–54, 55–64, and 65 years and above (see Fig. 6.1). The highest number of respondents (33%) lies in the 35–44 year category. If this category is combined with the 25–34 year category, a total of 55% of respondents were represented.

Md. F. Hossain and S. Momtaz, *Evaluating Climate Change Governance in Bangladesh*,
https://doi.org/10.1007/978-3-031-98486-0_6

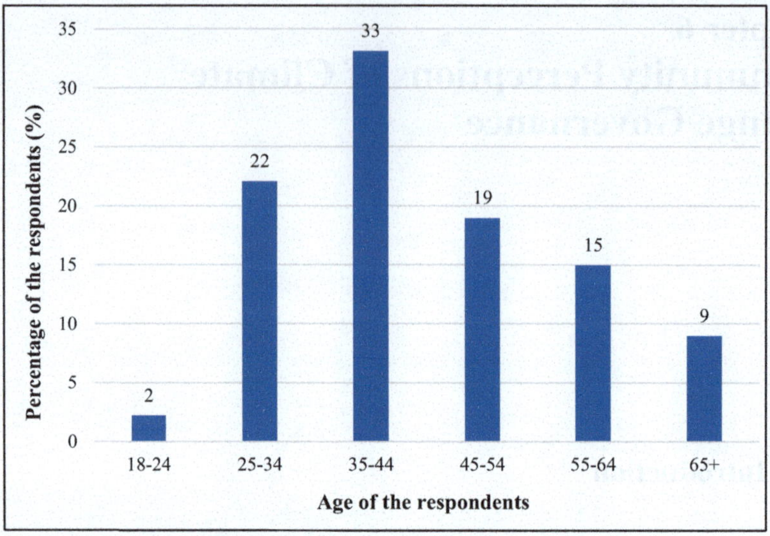

Fig. 6.1 Community (respondent) age structure. *Source* Fieldwork (2019)

Figure 6.1 shows that the majority of respondents are in the middle-age categories. These middle-aged respondents play an important role in several aspects of community life. Further, these respondents are old enough to recall previous climate change events and the measures taken at the community level during those times, which was a major consideration for the present study.

6.3 Educational Level

Education plays a vital role in a country's human resource development. Human development usually happens through building capability (UNDP, 2016). Education brings about considerable changes in societal behaviour (Baytiyeh, 2017). Education attainments also enable people to understand various aspects of disasters and climate events, and they can then stake measures to address these adverse situations. Hence, education is often considered the most influential component at the community level for human resource development.

The education level of the respondents was classified into five categories: illiterate, primary, secondary, technical and vocational level, and college and university level. The results show that 10% of the respondents are illiterate (e.g. they cannot read, write or sign their names). More than half (52%) of respondents had completed primary-level education (up to class five) (see Fig. 6.2). Among the remaining respondents, 29% had participated in secondary-level education, which incorporates the Secondary School Certificate examination in Bangladesh. Only five percent of

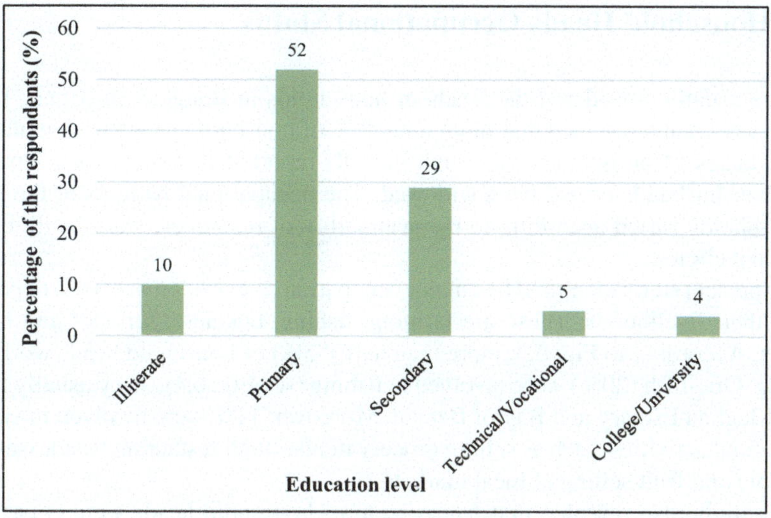

Fig. 6.2 Respondent education level. *Source* Fieldwork (2019)

respondents had completed technical- or vocational-level education and four percent had studied at college or university level (bachelor or masters programs).

A school teacher from Char Latia village of Char Iswar Union noted that:

> In the past schooling was not satisfactory level at all in this area. However, the perceptions of the guardians changed for sending their children to school nowadays. Many children, to date, are going to school by facing various troubles. Moreover, the schools are located far away from our locality. That's why, the children are facing difficulties to attend to schools regularly. Nevertheless, we're trying to convince our children and putting effort to send them school regularly. (FGD2)

However, local people claimed that the number of educated people was increasing currently; these people in turn were contributing at various levels in the community (FGD2). They also revealed that GOs conducted more educational activities in urban areas than in rural areas.

6.4 Marital Status

Marriage contributes to social bonding and ensures responsibility in marriage partners. In Bangladesh, the majority of adults are married once (BBS, 2015). The respondents included individuals with different marital statuses, such as married and unmarried. The majority (92%) of respondents (heads of households) for this study were married. The remainder (eight percent) were unmarried or single.

6.5 Household Heads Occupational Status

Men are usually considered the heads of households in Bangladesh (Chen, 1986). The survey results revealed that more than 90% of household heads were male. The female heads of households (although not all) reported that they were separated from their husbands or had been widowed. The occupational patterns of the heads of households varied according to the availability of resources, work facilities and individual choices.

The present study identified five major occupation categories that people depended on for their livelihoods. These are farming, fishing, business, services and home-making. According to Fig. 6.3, more than half (55%) of household heads worked in farming. One-fifth (20%) were involved in fishing (see Fig. 6.4). They usually fished in the Meghna Estuary and Bay of Bengal. Moreover, 11% were involved in various commercial activities, such as selling grocery items, small restaurant businesses, and vegetable and fruit selling at local markets.

This study also revealed that five percent of household heads were engaged in various services, mostly in local-level GOs, NGOs and other private offices (see Fig. 6.3). Another five percent, mostly the female household heads, were considered homemakers. Another four percent mentioned that they were not involved in any type of 'occupation' activity because of their old age.

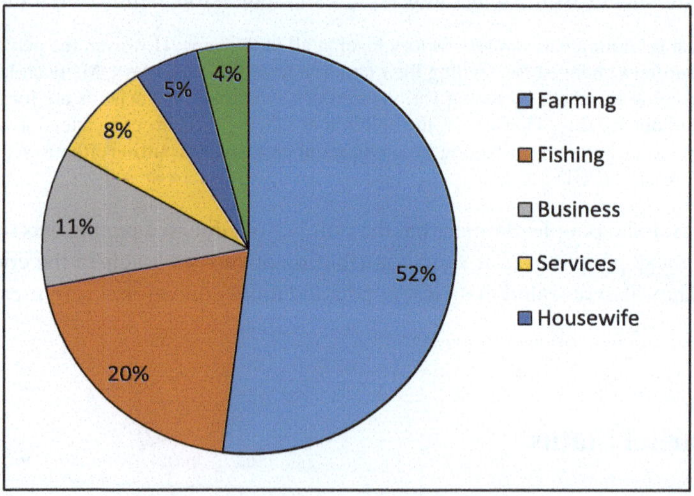

Fig. 6.3 Respondent occupation. *Source* Fieldwork (2019)

Fig. 6.4 Typical fishing boats and fishers in Hatiya. *Source* Fieldwork (2019)

6.6 Respondent House Types

The economic and social conditions of households determine housing patterns in Bangladesh. Moreover, local culture, environmental conditions and the availability of resources often influence traditional housing choices in Bangladesh (Ferdous et al., 2017). The fieldwork reveals that the majority of respondents (82%) were living in kutcha houses,[1] with corrugated iron sheets on their roofs and walls of mud and bamboo (see Fig. 6.6). Semi-pucca houses[2] were owned by 15% of respondents. About 3% of respondent houses were pucca,[3] with floor, walls and roofs of cement, bricks and iron rod (Fig. 6.5).

However, the selected respondents' observations provided an overview of the current climate change governance practices in the study area. They also explained their level of participation in climate change-related activities.

[1] A house with walls made of clay, wood, bamboo, straw or raw bricks and roofs made of tin, bamboo and straw is considered a kutcha house.

[2] Semi-pucca houses are those whose walls are generally made of cement and bricks with a roof made of tin, asbestos sheet, wood or bamboo.

[3] A house whose floor, walls and roof are made of cement, bricks and stones is called a pucca house (Bangladesh Bureau of Statistics, 2014).

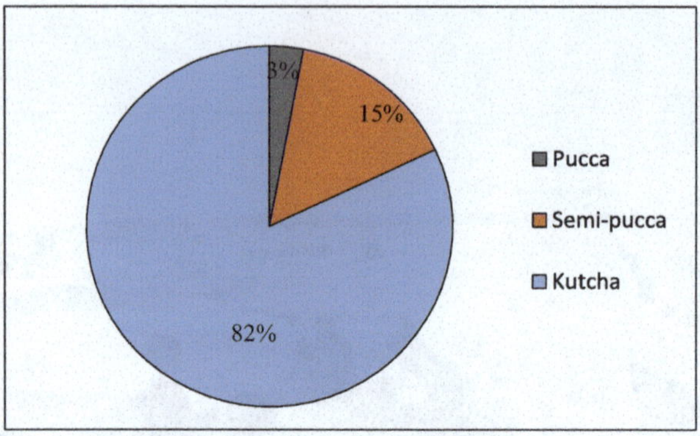

Fig. 6.5 Respondent house types. *Source* Fieldwork (2019)

Fig. 6.6 Typical kutcha house in Hatiya. *Source* Fieldwork (2019)

6.7 Respondents' Knowledge of Climate Change

Local people have their own understanding of climate change and this has been developed through their daily lives over a long period (Hossen, 2018). The interviewed respondents were asked for their perceptions about climate change in their respective areas. Most respondents (93%) said they had heard about climate change issues, as various programs and related activities had taken place in their area. Moreover, they considered themselves victims of climate change. Only seven percent of respondents claimed that they did not know about climate change issues in detail.

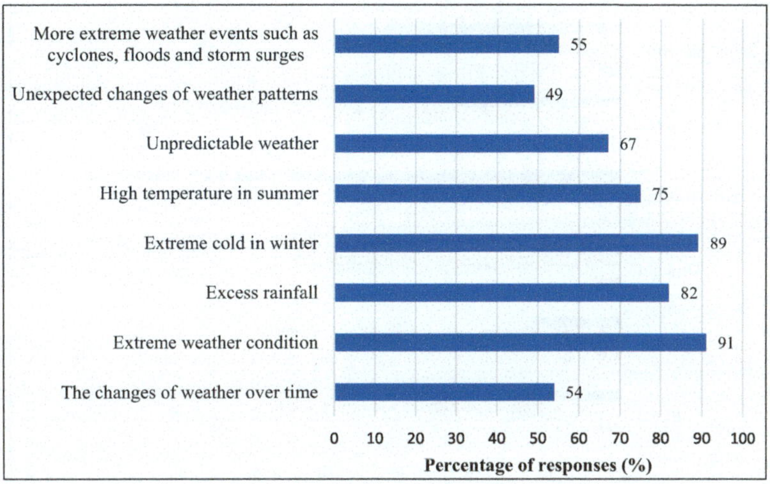

Fig. 6.7 Perceptions of climate change by respondents. *Source* Fieldwork (2019)

6.8 Perceptions of Climate Change

Most respondents had some idea about climate change issues. The respondents were asked how they perceived climate change on Hatiya Island. The majority (91%) mentioned climate change in these terms: ‘it is extreme weather condition’. Of the 181 respondents, 89% said that they noticed extreme cold in winter, whereas 75% observed higher temperatures in the summer. According to the locals, they suffered most significantly because of this temperature-related climatic variability. The survey results showed that a high number of respondents (82%) mentioned excess rainfall due to climate change in their areas (see Fig. 6.7). More than half (55%) of respondents described that extreme weather events such as cyclones, floods and storm surges were due to climate change, while 54% perceived changes in weather over time. ‘Unpredictable weather’ and ‘unexpected changes of weather patterns’ were mentioned by 67% and 49% of respondents, respectively.

6.9 Perception of Climate Events

The respondents were asked about the types of climate events they faced in their surrounding areas. Of the 181 respondents, 93% mentioned coastal erosion as ‘high’ (high + very high) in that area (see Fig. 6.9), whereas 57% reported drought as ‘high’. The survey results also showed that saline intrusion, storm surges, drought, cyclones and floods were indicated as ‘high’ by the respondents (70%, 62%, 57%, 64% and 81%, respectively). Thus, based on the Likert scale, climatic events with substantial effects on people are termed ‘high’. Events with less effect are characterised as

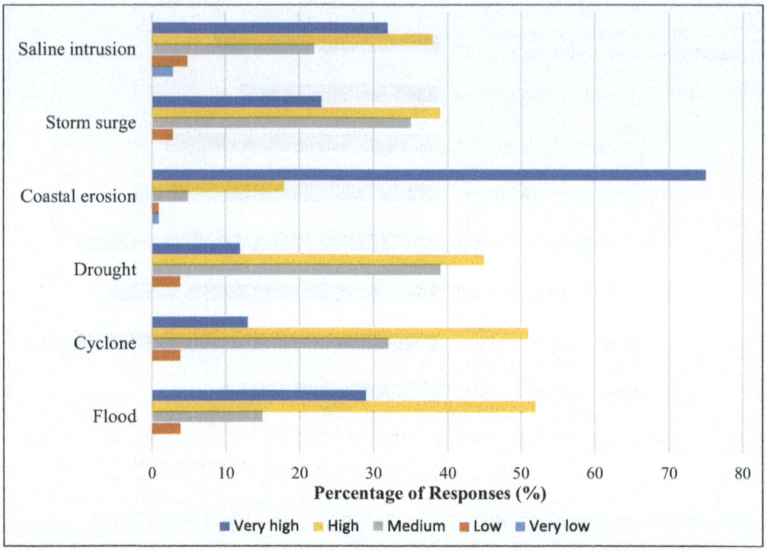

Fig. 6.8 Perceptions and level of climate events by respondents. *Source* Fieldwork (2019)

Fig. 6.9 Coastal erosion in Hatiya Island. *Source* Fieldwork (2019)

'medium', and those with minor effects on people are categorised as 'low' (Arévalo-Valenzuela et al., 2021). Many respondents reported saline intrusion (22%), storm surges (35%), drought (39%) and cyclones (32%) as medium-level events in their community. On the other hand, a few respondents nominated these events as 'low' (very low + low) (see Fig. 6.8).

Flood: 'Of course! Flood is the main problem in the monsoon. Our homes and lands are affected by floods in every full moon. We usually leave our houses when flood water appears.'

Cyclone: 'We were affected by cyclones several times in the past and thousands of people were affected and many of them died. Cyclone shelters are inadequate for this area. When cyclone starts, we feel really unsecured.'

River erosion: 'Our farmlands, homesteads, and the entire community areas were devoured by the sea within few days. In some years, the erosion was fast, while in other years the erosion rate was relatively slow. We've lost everything and passing a miserable life.'

Storm surge: 'Storm surge, in associated with floods, is observed in these areas. This also caused damages of crops, houses and public property.'

Drought: 'We usually get more water in monsoon but less in winter. Sometimes, our cultivation is often affected due to shortage of water in dry season. We also face difficulties to get drinking water from deep tube wells.'

Saline intrusion: 'Due to siltation in the river beds, the level of floodtide increased around two to three feet at the present time. As the embankment broke down in our area, saline water entered into our communities and local canals filled up with this water. This saline water caused substantial damages in the area. Moreover, crop production declined in this area dramatically and we hardly get one crop in a year.'

Lightning: 'In Hatiya Island, lightning events have increased in the last few years, and few people have died already. In fact, when lightning starts, we become scared and try to take shelter in a safe place. This appears like a new threat for us.'

Fig. 6.10 Responses to climate events. *Source* Fieldwork (2019)

The participants were asked what climate events they had experienced and how they had been affected in the last 10 years. Some typical responses are captured in Fig. 6.10.

6.10 Perceptions of Climate Change Effects

Climate change vulnerabilities and effects depend on a community's spatial, temporal scale and socioeconomic conditions (MoEFCC, 2009a, 2009b). Climate change effects are placing people's lives and livelihoods in danger, especially among poor and marginalised groups (UNDP, 2016). Respondents, faced with difficulties due to climate change, were asked about the effects of climate change-related events such as floods, cyclones, river and coastal erosion on their local environment, lives and livelihoods. They were given 12 statements and asked to provide their views (see Fig. 6.11).

However, perceptions were highest with respect to increases in natural disasters (91%), disease (90%), property damage (89%), poverty (86%) and decreases in work opportunities (89%), and income (87%). They perceived that the education system, which was impeded in these areas because of climate change-related events, caused suffering among children; 85% of respondents reported this.

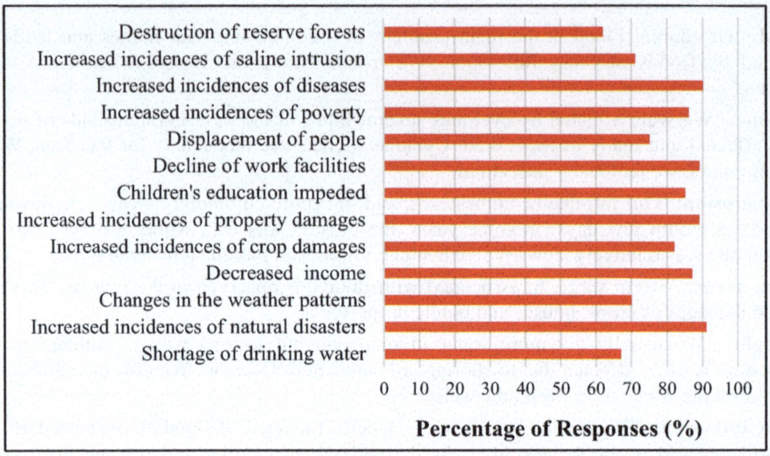

Fig. 6.11 Perceptions of respondents on the effects of climatic events. *Source* Fieldwork (2019)

Concern for the displacement of people in the study area was supported by 75% of respondents. Increased incidences of saline water intrusion and shortage of drinking water were reported among the communities, and these were mentioned by 71% and 67% of respondents, respectively. The local reserve forests were damaged over time by coastal and river erosion; this was indicated by 62% of respondents (see Fig. 6.12). Moreover, respondents reported that areas outside their embankment became flooded first during the monsoon. This upazila was severely affected by major cyclones in 1970 and 1991, with causalities numbering 35,000 and 5000, respectively (KII13, KII18, KII20). These caused major damage to crops and property (FGD1, FGD2, FGD3, FGD4). Further, saline-affected water entered the community through sluice gates and fresh water was difficult to obtain (KII20, FGD3).

6.11 Relationship of Stakeholders in Climate Change Governance

The relationships of stakeholders refer to how they work and practise climate change governance-related issues in their areas. Various stakeholders are working on different aspects of climate change governance issues, leading them to prioritise their own policies and rules. As a result, their preferred climate change governance systems also differed. Based on the Likert scale with a range from 1 to 5 (e.g. 1 = very good, 2 = good, 3 = medium, 4 = bad, and 5 = very bad), choices were considered while conducting the household survey (Bañas et al., 2020; Cen-López & Aguilar-Perera, 2020). According to the respondents, more than half (53%) perceived that the present relationships of stakeholders are at a medium level (i.e. their relationships are neither good nor bad). Of 181 respondents, 29% noted the relationships as bad

Fig. 6.12 Destruction of reserve forest due to coastal erosion in Hatiya Island. *Source* Fieldwork (2019)

(very bad + bad), as they did not maintain proper relationships as expected. In addition, 18% were observed as good (good + very good); they were trying to maintain esteemed relationships with relevant stakeholders (see Table 6.1).

According to community perceptions (see Table 6.2), government officials maintained medium-level relationships, with a WAI of 0.41. They indicated that various government departments working at the community level are working on climate change issues, but were unable to meet the expected relationship outcomes. Community members also acknowledged that government departments were working on various aspects to reduce the effects of climate change. Nevertheless, in terms of relationships, government officials scored the lowest (0.41) among the mentioned four stakeholders (see Fig. 6.12).

In contrast, community members perceived that NGO officials maintained strong relationships with other stakeholders, with a WAI of 0.66. According to respondents, NGOs were more dynamic in their activities; the community received several services from NGOs during and after climate change-related events. NGOs can reach communities more easily than other stakeholders. However, CBOs and community members were maintaining relationships with other stakeholders: with WAI results of

Table 6.1 Respondents' stakeholder relationship ratings

N = 181				
Responses (%)				
Very bad	Bad	Medium	Good	Very good
2	27	53	15	3

Table 6.2 Stakeholder relationships in climate change governance

Name of stakeholders	Very weak	Weak	Medium	Strong	Very strong	N = 181 WAI	OA
Governmental organisation officials	10	46	17	25	2	0.41	M
Non-governmental organisation officials	2	11	15	64	8	0.66	S
Community-based organisations	4	33	35	24	4	0.48	M
Community members	6	36	27	29	2	0.46	M

Note Very weak (VW): 0.01–0.2; weak (W): 0.21–0.4, medium (M): 0.41–0.6; strong (S): 0.61–0.8; very strong (VS): 0.81–1. *WAI* weighted average index, *OA* overall assessment

0.48 and 0.46 respectively (see Table 6.2 and Fig. 6.13). CBOs are performing better in maintaining relationships than community members (0.66 vs. 0.46), because they have active institutional settings at the community level, enabling them to manage climate change governance practices.

A high-level employee of a national NGO indicates how they coordinate with the local administration on climate change-related activities:

> We often contact with [the] local upazila administration and Union Council authority for facilitating field-level activities. In the monthly meeting at upazila level, our staff raise various issues relating to field-level implementation. If we face difficulties with local Union

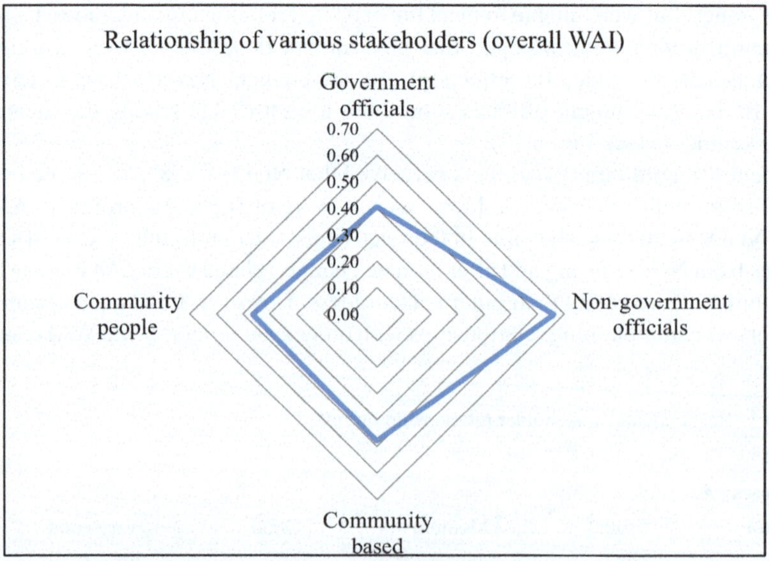

Fig. 6.13 Overall WAI by stakeholder relationship. *Source* Fieldwork (2019)

Council authority, we try to solve the issue by discussing with upazila administration. We often seek the approval from the local administration prior to start the field-level activities. For example, if we want to excavate the land, we need the permission from the UNO as it is linked with the degradation of natural resources. Sometimes we get the approval very easily, and sometimes the issues become complicated and cause delay. (KII09)

Therefore, the anticipated relationship of various stakeholders involved in climate change-related activities at the local level is vital to gain the expected benefits from projects and programs. However, the overall relationship of stakeholders in the climate change governance issue is considered 'good'. Improved relationships may help to overcome challenges that may arise during project implementation phases.

6.12 Climate Change Governance Practices at the Local Level

Effective climate change governance practices are important to achieve better results from climate change-related projects at the local level. Climate change governance practices are different at the local (subdistrict) level than those at the national level. Most activities related to climate change are managed and implemented by the upazila (subdistrict) administration headed by the UNO. At the local level, many issues are considered based on the local context. Five criteria—accountability, transparency, participation, government effectiveness and resources—were chosen in this study to evaluate climate change governance practices. Respondents were asked to provide their levels of agreement and disagreement against several statements related to these criteria.

6.12.1 Accountability

Accountability in climate change governance means that the relevant actors are responsible for their activities (Biermann & Gupta, 2011). Accountability is an important criterion in climate change governance (Biermann et al., 2010; Mees & Driessen, 2018), helping to provide better outcomes from climate change-related activities when compared with existing governance systems (Biermann et al., 2010). Nevertheless, this was often lacking in projects relating to climate change at the local level. These projects were mainly conducted by local contractors, resulting in them focusing on their own interests rather than maintaining quality in their work. Thus, people were not getting the expected benefits because of the absence of proper implementation strategies in government projects (KII18, KII20).

Table 6.3 illustrates the WAI and OA results regarding the accountability levels of climate change-related activities. Local people sometimes visited these projects in their areas to see the progress of work; and they deemed the OA value as weak (WAI = 0.31). Moreover, locals' impressions emerged from visits to both GO and NGO

Table 6.3 Perceptions of respondents' on accountability in climate change-related activities

Issues	WAI	OA
Locals' visits to climate projects	0.31	W
Locals' visits to governmental organisation (GO) offices	0.26	W
Locals' visits to non-governmental organisation (NGO) offices	0.38	W
GO officials' visits to climate projects	0.31	W
NGO officials' visits to climate projects	0.48	M

Note Very weak (VW): 0.01–0.2; weak (W): 0.21–0.4, medium (M): 0.41–0.6; strong (S): 0.61–0.8; very strong (VS): 0.81–1. *WAI* weighted average index, *OA* overall assessment

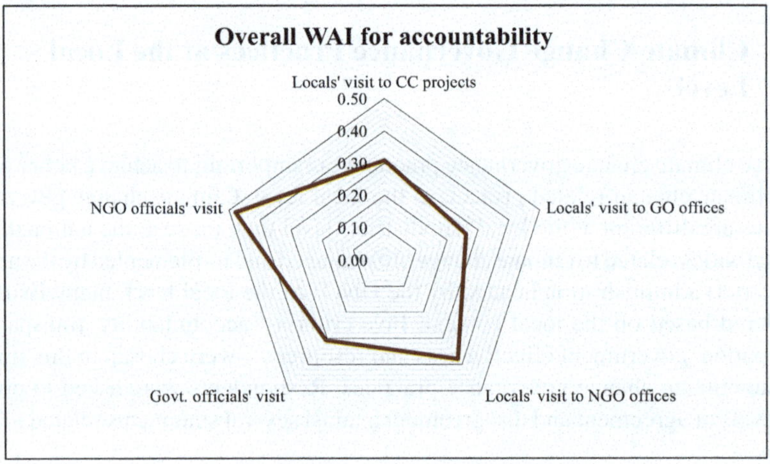

Fig. 6.14 Respondents' perceptions of accountability. *Source* Fieldwork (2019)

offices to observe the status of work. The OA values for both were weak; the WAI values were 0.26 and 0.38 respectively. Similarly, one indicator (e.g. 'government officials' visit') was also weak (WAI = 0.31), while another (e.g. 'NGO officials' visit') was strong (WAI = 0.48). This shows a significant difference in WAI (0.31 vs. 0.48) between the GO and NGO official visits at the local level (see Fig. 6.14). NGO officials visited their climate project sites more compared with GO officials. Moreover, local people perceived that the existing climate change governance practices in terms of accountability were unlikely to bring significant change to locals' lives and livelihoods.

6.12.2 Transparency

Transparency is becoming an important criterion within public and private climate governance activities; it enhances accountability and increases environmental outcomes (Gupta & Mason, 2016). As a mid-level government officer explains:

> Every government program has some specific formats and local government authorities follow those formats to implement various climate change-related projects and programs. (KII15)

At the beginning of a project, the local administration receives work orders from the union administration relating to climate change-induced disasters (KII19). Then the upazila administration selects work on a priority basis and takes the necessary steps to implement those works: from upazila to union and then to village level (KII14, KII19).

In the colonial period,[4] the local government made demands of the central government, and this situation persists in Bangladesh. There is no transparency and no inclusiveness in the activities of local administrations on climate change issues (KII22).

The respondents were asked to indicate their perceptions about transparency. To assess this, three issues were considered: support for climate victims, access to information, and timeliness (Barua et al., 2014). Based on a Likert scale (Bañas et al., 2020; Cen-López & Aguilar-Perera, 2020), more than half (55%) of respondents perceived that the 'level of effort to facilitate the climate victims' issues was 'bad' (see Fig. 6.14). Around one-fourth of respondents nominated 'medium', and 22% noted it as 'good'. In case of 'access to information', around half (44%) of respondents nominated it as 'bad', while 38% indicated it was 'good'. When the respondents were questioned about the 'timeliness' of climate change-related activities, more than half (57%) of noted it was 'bad'. In contrast, only one-fourth (25%) of respondents perceived it as 'good'. More than half of the respondents noted that transparency was not adhered to properly at the local level. A local journalist explains this:

> Actually, we don't see to follow transparency issue in climate change-related projects and programs in our areas. The contractors do whatever they want and they don't maintain the quality of works. The allocated funds aren't used properly due to the misdeeds of various government organisations and political leaders. (KII17)

6.12.3 Government Effectiveness

In the case of climate change governance, government effectiveness is described as the institutional capacity to formulate and implement policies focusing on climate change effectively (Kaufmann et al., 2011). Government effectiveness is an important aspect affecting governance in a particular area. The local administration usually

---

[4] The colonial period in Bengal was from 1757 to 1947. The English East India Company ruled over Bengal during this time.

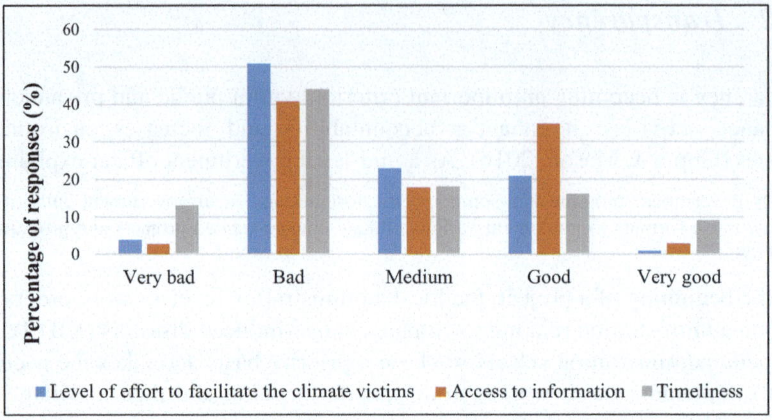

Fig. 6.15 Respondents' perceptions of transparency. *Source* Fieldwork (2019)

follows national-level policies and implementation strategies on climate change when implementing projects. They mainly follow instructions from the central government to ensure implementation at the field level. They also follow some generic responsibilities that address climate change, such as organising campaigns and awareness-raising programs for the community (GoB, 2012). Respondents were asked to state their perceptions of government effectiveness in the study area. More than half (54%) of the respondents (strongly disagree + disagree) revealed that projects relating to climate change did not address the community's problems. However, one-third (33%) of respondents perceived that climate change-related activities were affected by political pressure (see Fig. 6.15).

However, approximately half (43%) of respondents indicated that climate change projects were adopted in the study area, while 39% disagreed with this. More than half (56%) of respondents noted that climate change projects were implemented properly in the area through adhering to policies and implementation strategies, whereas one-third (33%) disagreed with this. Further, 62% of respondents disagreed about 'the reliability of government's commitment to policies', while a few (18%) supported it (Table 6.4).

6.12.4 *Resources*

UNO, the chief executive officer at the subdistrict level, mentioned that the upazila administration was trying to support a large number of climate victims living in the area with limited government allocation. The UNO of Hatiya Upazila explains the scarcity of resources:

Table 6.4 Perceived views of locals on government effectiveness in climate change-related activities

Issues	Responses					WAI	OA
	Strongly disagree (%)	Disagree (%)	Neither agree nor disagree (%)	Agree (%)	Strongly agree (%)		
Climate change-related projects addressed the problems of community people	11	43	13	23	10	0.45	M
Climate change projects were affected by political pressure	7	32	28	31	12	0.57	M
Proper policy and implementation of climate projects	10	46	15	25	4	0.42	M
The reliability of government commitment to policies	14	48	20	15	3	0.36	W

Source Fieldwork (2019)
Note Very weak (VW): 0.01–0.2; weak (W): 0.21–0.4, medium (M): 0.41–0.6; strong (S): 0.61–0.8; very strong (VS): 0.81–1. *WAI* weighted average index, *OA* overall assessment

> If we consider government's priority, then the allocations for managing the effects of climate change were adequate, but these were insufficient for the recipients. However, upazila administration couldn't implement various projects focusing on climate change issue due to insufficient funds. We receive funds from the government departments, donors and NGOs. These funds are distributed to the Union Council authorities also. (KII19)

The Union Council authorities could not perform their task properly because of corruption, partisan politics and nepotism (KII13, KII17). As a result, vulnerable people would often not receive benefits from these types of government allocation. The UNO also demonstrates resource scarcity at the local level:

> If upazila administration is allocated 100 blankets to provide the disaster affected people, usually more than 1,000 people gather to get their blankets. So, one out of ten aspirants can receive one blanket. The people who aren't enlisted for the blankets and couldn't obtain blankets, they become angry and blame the government departments for this. (KII19)

However, local communities were asked about the availability of resources for work on climate change issues. Around half (48%) of the respondents neither agreed nor disagreed: they rated the issue at a 'neutral' level (see Fig. 6.16). Among the

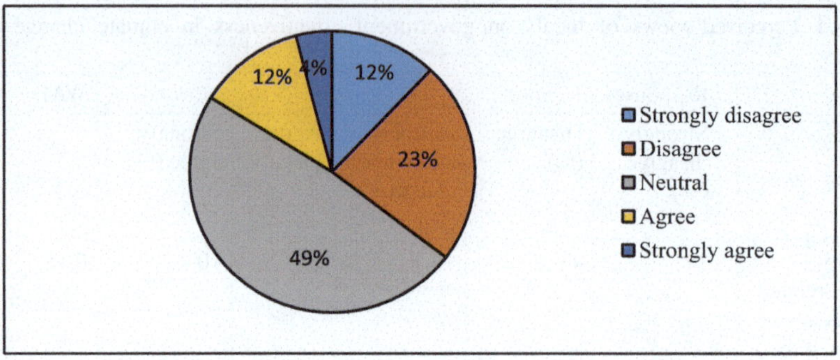

Fig. 6.16 Respondents' perceptions of resource availability. *Source* Fieldwork (2019)

respondents, 36% disagreed, whereas 16% agreed that the available resources could be used properly to make the existing situation better. This situation is explained in the following extracts:

> We come to know from various sources that a good amount of funds are being allocated to Bangladesh but we don't see any presence of these funds in our areas. So we're in doubt about the destination and effectiveness of these funds. (FGD1)

> We aren't getting required support to tackle the climate change issue. We need more funds to tackle the impacts of climate change at the community level in this area. If we can make the best uses of available resources in climate change-related climate change activities, then it can bring better outcomes for the communities. That's why, the effective climate change governance is an unavoidable issue at the local level also. (KII14)

Allocated resources are not enough to implement projects and programs relating to climate change, with significant outcomes at the local level (KII19). Moreover, available funds are not being used properly due the poor performance of local administrations, political interference and misused funds at several levels of implementation (KII17, KII18, KII20).

6.12.5 Participation

Climate change governance is based on the participation of various stakeholders, including local administrations, NGOs, CSOs, and members of the public (Sarzynski, 2015; Sprain, 2016). Participation will result in more effective methods at the national, regional and international levels for climate change governance (Jodoin et al., 2015; Kabiri, 2016). This could bring about better outcomes at the community level when projects are implemented. Upazila administration, following the government format, incorporates stakeholders from different levels to implement climate change-related works effectively. However, the participation of community representatives is vital to ensure the quality of work. The capacity of local GOs is not enough to deal with multi-disciplinary climate change issues (KII19) (KII23). UNO, the head of local administration in Hatiya Upazila describes this participation issue:

> We actually try to incorporate various stakeholders such as government officials, NGO officials, local political leaders, community-based organisation (CBO) representatives, and others in various programs relating to climate change. We, however, in many cases, can't do accordingly due to misuses of property and poor capacity of people to receive various facilities. Of course, with the available resources, we can get better results by implementing various projects properly. (KII19)

> We always incorporate local-level representatives such as Union Council chairmen, members, and they represent some of the specific communities. We also invite few NGOs those are working in the grassroot level. Apart from this, we also invite few farmers. So, we try to listen from various groups and their problems, and their feedback to overcome the issues. (KII19)

Field survey results show that around half (48%) of the respondents perceived local participation as weak (very weak + weak) (see Table 6.5). More than one-third (37%) of respondents thought local participation was 'medium', and only 15% indicated it was 'strong'. More than half (54%) of the respondents mentioned that the involvement of stakeholders in climate change-related activities was 'very weak'. Moreover, more than one-third (37%) of respondents perceived it as 'very strong'. WAI values for the participation of locals and various stakeholders are 0.36 and 0.45, successively (see Table 6.5). So, respondents demonstrated that stakeholders, apart from locals, participated more in climate change-related activities.

Table 6.5 Perceived views of locals on participation in climate change-related activities

Statements	Responses					WAI	OA
	Very weak (%)	Weak (%)	Medium (%)	Strong (%)	Very strong (%)		
Participation of community people in climate change governance practices in terms of engagement and contribution	28	20	37	12	3	0.36	W
Involvement of stakeholders in climate change-related activities, such as in planning and implementation	8	46	9	33	4	0.45	M
Total respondents	181						

Source Fieldwork (2019)
Note Very weak (VW): 0.01–0.2; weak (W): 0.21–0.4, medium (M): 0.41–0.6; strong (S): 0.61–0.8; very strong (VS): 0.81–1. *WAI* weighted average index, *OA* overall assessment

In summary, community member participation in climate change governance practices at the local level does not operate at the expected level (WAI 0.36). Nevertheless, the OA relating to the involvement of various stakeholders in climate change-related activities (i.e. planning and implementation) is comparatively good (WAI 0.45) (see Table 6.5).

6.13 Overall Climate Change Governance Status at the Local Level

Figure 6.17 shows the overall WAI results on climate change governance in the study area. For transparency, the OA result was rated as weak (WAI = 0.35), which was derived from the data in Fig. 6.14 (i.e. transparency was not maintained properly in various activities relating to climate change activities). Conversely, with respect to accountability, government effectiveness and resources, these three criteria were comparatively better followed in the study area. The WAI results are 0.44, 0.45 and 0.43, respectively (see Table 6.6). WAI values for accountability, government effectiveness and resources are calculated from Tables 6.3, 6.4 and Fig. 6.16, respectively. According to the respondents, these three criteria were improving gradually and making a significant positive impact on locals' lives and livelihoods.

Being a part of various activities relating to climate change, different stakeholders did not participate in local communities as much as expected (KII18, KII20). The OA result shows that participation at the community level was weak. The WAI value for participation was 0.40 (calculated from Table 6.5). So, the overall WAI results

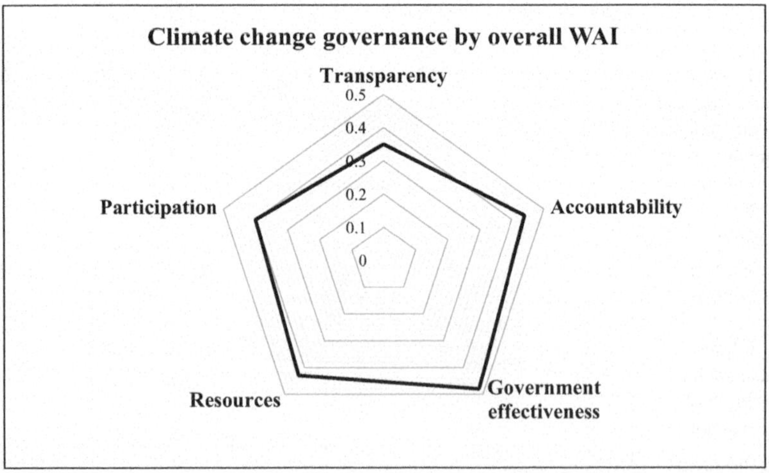

Fig. 6.17 Climate change governance status at the local level by overall WAI

Table 6.6 Climate change governance status by WAI and OA results

Criteria	Total WAI	OA
Transparency	0.35	W
Accountability	0.44	M
Government effectiveness	0.45	M
Resources	0.43	M
Participation	0.40	W

Note Very weak (VW): 0.01–0.2; weak (W): 0.21–0.4, medium (M): 0.41–0.6; strong (S): 0.61–0.8; very strong (VS): 0.81–1. *WAI* weighted average index, *OA* overall assessment

demonstrate that the obtained values for 'accountability', 'government effectiveness' and 'resources' are better than those for 'transparency' and 'participation' (see Fig. 6.17).

Local communities manage disaster events in their own manner, manage disasters in a specific way; this is necessary as responses and suffering is unique to each community. They select seeds for crop production and follow suitable agricultural practices; they prepare for cyclones and storm surges. An expert explains the concerns regarding climate change at the community level:

> We're keen to provide reliefs during the disaster and post-disaster period. If we go through disaster risk reduction (DRR) principals or resolve climate change adaptation issue in planning stage, and this realisation is here but not started yet. (KII22)

6.14 Perceptions About the Effectiveness of Climate Change Governance

In all, 98% of respondents agreed that they were affected by climate change-related events. They were asked about their perceptions on the effectiveness of climate change governance. They were provided with 25 statements and asked to rank their levels of agreement or disagreement (see Table 6.7).

The results from the field survey showed that at least 50% of responses supported the effectiveness of climate change governance in 12 out of 25 issues. However, perceptions were highest for 'improved climate change governance is expected' (88%), 'climate projects are enhancing public awareness' (85%), 'female members' participation [is] encouraged in various projects' (75%) and 'role plays in overall development' (72%).

In contrast, lower level responses were reported for different issues such as 'proper utilisation of climate funds' (11%), 'timely completion of climate projects' (15%), 'regular monitoring of climate projects' (22%), 'adequate fund allocation for climate projects' (25%), and 'institutional capacity building' (27%).

Table 6.7 Effectiveness of climate change governance as perceived by respondents

Issues	Responses				
	Agree strongly (%)	Agree (%)	Neither agree nor disagree (%)	Disagree (%)	Disagree strongly (%)
Presence of institutional setting for climate governance	7	52	3	29	9
Presence of adequate plans and programs	4	58	2	26	10
Policy and plans are in favour of people	5	48	4	31	12
Participation of various stakeholders are ensured	3	53	3	31	10
Climate change projects ensures the involvement of locals	3	57	3	29	8
Climate change projects involve locals in decision-making processes	4	56	2	25	13
Existing climate change projects are supportive for local livelihoods	3	50	7	26	14
Female members' participation is encouraged in various projects	5	70	1	21	3
Role plays in overall development	7	65	4	17	7
Climate change projects are enhancing public awareness	6	79	3	11	1
Climate change governance has changed over time	4	62	10	13	11
Improved climate change governance is expected	20	68	2	7	3
Personal rights and responsibilities in climate governance practices are not well defined	5	28	8	34	25
Adequate fund allocation for climate projects	16	9	0	30	45
Proper use of climate funds	4	7	0	36	53
Access to information on climate change projects	2	43	3	37	15
Timely completion of climate projects	2	13	3	41	41
Regular monitoring of climate projects	3	19	2	41	35
Climate change governance system is effective	5	37	3	23	32
Rules and policies are well informed	3	41	6	43	7

(continued)

Table 6.7 (continued)

Issues	Responses				
	Agree strongly (%)	Agree (%)	Neither agree nor disagree (%)	Disagree (%)	Disagree strongly (%)
Employment opportunities for locals have not been created	10	25	3	27	35
Vulnerability and uncertainty of livelihood have not decreased	10	26	4	22	38
Confrontations are observed among government and non-government organisations in climate change-related projects	2	43	21	18	16
Stakeholder capacity building	4	35	15	33	13
Institutional capacity building	0	27	13	46	14

Source Fieldwork (2019)

In relation to 11 of the 25 issues in this study, at least 50% of responses disagreed on the effectiveness of climate change governance (see Table 6.7). Nevertheless, perceptions were highest with respect to 'proper utilisation of climate funds' (89%), 'timely completion of climate projects' (82%), 'regular monitoring of climate projects' (76%), 'adequate fund allocation for climate projects' (75%) and 'employment opportunities for locals have not been created' (62%). Conversely, a very low number of responses supported the hypothetical that 'improved climate change governance is expected' (10%), 'climate projects were enhancing public awareness' (12%), 'female members' participation are encouraged in various projects' (24%), 'role plays in overall development' (24%) and 'climate governance has changed over time' (24%).

There were some issues with no significant agreement or disagreement, such as 'confrontations are observed among GOs and NGOs in climate change-related projects' (21%), 'stakeholders' capacity building' (15%) and 'institutional capacity building' (13%).

6.15 Challenges for Local Climate Change Governance

At the local level, challenges are observed that are beyond the capacity of the community to respond to (Keskitalo & Kulyasova, 2016). Hatiya Island, as an area vulnerable to multiple disasters, has been affected by climate disasters. Climate change governance is important at the local level to meet the challenges of climate change-related issues. In Hatiya, the upazila administration has formed committees to work on disaster events from the upazila to the union level. They have had some experience tackling climate events, but climate change has not been a great focus of their actions (see Table 6.8).

The local administration has a low level of understanding about climate change issues. Local government staffs, however, were assigned to conduct activities relating to climate change. They were not skilled enough and their low capacity has often placed them in challenging situations when seeking to enhance community resilience to climate change (Vivekanada, 2013). As a result, they do not implement projects properly and the expected results are not obtained. For example, the community suffered much from the damaged road networks in the area (see Fig. 6.18). A high-level employee from a reputed NGO indicates the of local administrations' limitations:

> The local authorities don't consider or integrate the knowledge of local people which is very important for the successful completion of projects. They don't consider the local context. For example, if the local authority builds a cyclone shelter close to a river which may be eroded within a few years, and these types of works aren't considered as community supportive efforts at all. (KII26)

Stakeholder participation is lacking in many projects relating to climate change issues, but this is vital to ensure effective outputs from those projects (KII18, KII20). Local administration officials often do not consider the community suggestions for improving the effectiveness of projects (KII17, KII20, FGD1, FGD3). Moreover, the community expects more from the current local administration, but they often cannot reach administrative staff to convey their concerns.

Table 6.8 Challenges for local climate change governance

Class types	Challenges
Economic	Inadequate funds for climate projects (KII13, KII14, KII19) Climate change issues receive less focus (KII13, KII18, FGD3) Lack of trained people (KII17, KII18, FGD4)
Informational	Lack of expertise (KII13, KII20) Public participation is not always maintained (KII17, KII20) Minimal monitoring and evaluation (KII17, KII18, FGD1) Little access to information (KII13, KII17)
Institutional	Dependency on national government (KII15, KII19) Lack of effective local governance (KII14, KII16) Lack of coordination (KII18, KII20, FGD2) Less effort to incorporate into a structured (KII15, KII18) Lack of cooperation (KII16, KII20) Weak regulatory framework (KII18, KII20, FGD4) Less control over utilities (KII16, KII20, FGD3)
Political/cultural	Lack of climate policy for the local level (KII13, KII20, FGD4) Local political pressure (KII17, KII18, KII20) Less prioritised issue (KII14, KII17, KII20)

Source Fieldwork (2019)

Fig. 6.18 Damaged village roads in the study area. *Source* Fieldwork (2019)

The challenges in local climate governance have been identified by the present study. According to Sippel and Jenssen (2009), these can be classified as 'economic', 'informational', 'institutional' and 'political/cultural' (see Table 6.8).

The government and other agencies are giving funds to implement various projects relating to climate events in this area. According to the KIs, however, these are inadequate for the project requirements. As a result, projects gaining approval do not meet the community's expectations. Local political pressure groups, government departments and other beneficiaries are not making sufficient effort for the proper implementation of these projects (KII17, KII29). A faculty member from a government college of Hatiya Island notes that:

> If we see the success rates in various climate change-related projects, which aren't expected level. The contractors, assigned for developing climate projects, need to manage various groups to implement their works. Moreover, various projects were shifted by the government departments from this area to other areas due to conflicts of several local groups in the past. When local influential groups receive work orders, there are chances to deteriorate the quality of works considerably. Apart from this, they manage the monitoring agencies without maintaining the standard of works. (KII20)

Local government departments dealing with DM, forests, fisheries, LGED and agriculture are working towards implementing programs relating to climate change and complying with existing policies. For example, local forests have been declining (see Fig. 6.19) (KII18, KII19). The FD, working at the local level, was responsible for managing the forests in Hatiya Island (KII15). When they started afforestation programs in a particular area, after few days influential local groups occupied these areas through leases and with the tacit approval of the local administration (KII15). Later, when granted a lease, they cleared the forests for cultivation, housing and other purposes (KII17, KII18). Thus, a particular group benefited from destroying reserve

Fig. 6.19 Illegal logging in the reserve forest of Hatiya. *Source* Fieldwork (2019)

forests and left islanders in a vulnerable position regarding climate change-induced disasters (KII18, KII19).

Public awareness of climate change issues is also important at the local level; however, local people have not been aware enough to deal with specific climate change issues (KII19). Moreover, they cannot ensure their rights at the community level regarding various projects relating to climate change (KII18). Consequently, they have not been seeing the expected results from projects implemented by GOs and NGOs (KII13, KII16).

Although the national government has already undertaken numerous measures to tackle climate change, in many cases, the results are not reflected at the community level as expected. This is evident in the following extracts:

> All the development activities are centrally driven and the local government institutions don't have enough capacity to deal climate change-related issues. (KII22)
>
> Limited capacity of government officials in local-level organisations can't bring considerable results, specially the analytical, planning, monitoring and evaluation. (KII23)
>
> Some people are getting support from all and some aren't getting at all. These things need to be addressed, otherwise it would be difficult to ensure the climate change governance practices at the local level. People have various necessities and they're really in crisis. (KII12)

Moreover, community perceptions and views have not always been considered when planning and implementing projects related to climate change at the local level (KII20, FGD2). If programs relating to climate change are implemented properly with adequate consideration of local knowledge, these programs could likely bring about significant changes to the community resilience required to face climate change effects. Therefore, the proper implementation of various climate change-related projects by GOs and NGOs is important to enhance local community resilience.

6.16 Discussion and Conclusion

Various local organisations in the form of public, private and civic bodies make efforts to assess the effects of climate change at the community level (Rahman & Huang, 2019). The communities of Hatiya Island are highly vulnerable to climate change; more so than the mainland. This area has been affected by severe climatic events in the past and the number of these events has increased, causing enormous damage to lives and property. Thus, it is vital to ensure effective climate change governance practices for the proper implementation of projects and programs addressing climate change at the local level.

Climate change governance criteria, such as accountability, transparency, participation, government effectiveness and resources, are key to providing effective responses to climate change-related issues that meet the expectations of local communities. The local administration has tried to maintain these criteria while operating climate change-related activities. However, in many cases, they cannot manage projects properly owing to a shortage of funds, and various limitations such as the lack of skilled staff and required political support at the local level. In addition, accountability is not well developed among stakeholders at the local level. Transparency in climate change-related projects is necessary because this helps maintain accountability and participation. This issue is followed in many projects and programs at the local level, but more possibilities must be considered at various levels for the successful completion of climate change-related projects.

Sometimes, local authorities do not consider or integrate the knowledge of local people; this is vital for the successful completion of locally led climate projects. At times, they do not examine the local context, culture, dynamics or needs to ensure the effectiveness of climate change-related projects. Local-level organisations cannot differentiate between climate finance and development finance when funds are channelled into various projects; this creates challenges for other relevant stakeholders.

According to the surveyed respondents, NGOs are performing better than government departments at the local level. Therefore, the capacity of local-level GOs must be strengthened to properly implement climate change-related projects. The community will then also benefit more from these projects. Despite the various challenges that are common at the local level, climate change governance practices are considered 'moderate', as based on the perceptions of locals. However, a combined effort, especially by GOs and NGOs at the community level, may reduce climate change's effects on lives and livelihoods substantially. For this to happen, all relevant stakeholders must come forward and make a collective effort to tackle climate change in this area.

Climate change is gaining importance for stakeholders at the local level, because of the significant and increasing exposure to climate change-induced disasters: daily life is becoming more difficult. Community members are facing the adverse effects of climate change because they are not yet prepared to tackle the issues properly. However, the community is able to perceive climate change issues better at present, and can undertake measures themselves, reducing adverse effects through effective climate change governance practices at the local level.

Finally, Chap. 7 presents the concluding remarks on the previous chapters, and a brief summary of the present study on climate governance perspectives in LDCs such as Bangladesh, with some recommendations and possible future research.

References

Arévalo-Valenzuela, P., Peña-Cortés, F., & Pincheira-Ulbrich, J. (2021). Ecosystem services and uses of dune systems of the coast of the Araucanía Region, Chile: A perception study. *Ocean & Coastal Management, 200*, 105450. https://doi.org/10.1016/j.ocecoaman.2020.105450

Bañas, J. C. C., Subade, R. F., Salaum, D. N., & Posa, C. T. (2020). Valuing vanishing coasts: The case of Miagao coastline in Southern Iloilo, Philippines. *Ocean & Coastal Management, 184*, 105008. https://doi.org/10.1016/j.ocecoaman.2019.105008

Bangladesh Bureau of Statistics. (2014). *Rural credit survey 2014*. www.bbs.gov.bd

Bangladesh Bureau of Statistics. (2015). *Trends, patterns and determinants of marriage in Bangladesh*. http://203.112.218.65:8008/WebTestApplication/userfiles/Image/PopMonographs/Volume-13_MM.pdf

Barua, P., Fransen, T., & Wood, D. (2014). *Climate policy implementation tracking framework*. Retrieved from World Resources Institute: https://www.wri.org/sites/default/files/climate_tracking_framework_working.pdf

Baytiyeh, H. (2017). Can disaster risk education reduce the impacts of recurring disasters on developing societies? *Education and Urban Society, 50*(3), 230–245. https://doi.org/10.1177/0013124517713111

Biermann, F., & Gupta, A. (2011). Accountability and legitimacy in earth system governance: A research framework. *Ecological Economics, 70*(11), 1856–1864. https://doi.org/10.1016/j.ecolecon.2011.04.008

Biermann, F., Betsill, M. M., Gupta, J., Kanie, N., Lebel, L., Liverman, D., Schroeder, H., Siebenhüner, B., & Zondervan, R. (2010). Earth system governance: A research framework. *International Environmental Agreements: Politics, Law and Economics, 10*(4), 277–298.

Cen-López, A., & Aguilar-Perera, A. (2020). Perceptions of diver-fishermen and recreational divers on lionfish as a threat in the Parque Nacional Arrecife Alacranes, southern Gulf of Mexico. *Ocean & Coastal Management, 193*, 105225. https://doi.org/10.1016/j.ocecoaman.2020.105225

Chen, M. (1986). Poverty, gender, and work in Bangladesh. *Economic and Political Weekly, 21*(5), 217–222.

Ferdous, L., Kafy, A., Gafur, A. K. R., & Wakil, M. A. (2017). An analysis on influencing factors of rural housing and settlement pattern in Rajshahi, Bangladesh. *Landscape Architecture and Regional Planning, 2*(4), 99–109. https://doi.org/10.11648/j.larp.20170204.12

Government of Bangladesh. (2012). *Bangladesh climate public expenditure and institutional review*. Government of the People's Republic of Bangladesh.

Gupta, A., & Mason, M. (2016). Disclosing or obscuring? The politics of transparency in global climate governance. *Current Opinion in Environmental Sustainability, 18*, 82–90. https://doi.org/10.1016/j.cosust.2015.11.004

Hossen, M. A. (2018, September 4). Social learning necessary for climate change adaptation. *Daily Observer*. http://www.observerbd.com/details.php?id=156482

Jodoin, S., Ducky, S., & Lofts, K. (2015). Public participation and climate governance: An introduction. *Review of European Community & International Environmental Law, 24*(2), 117–122.

Kabiri, N. (2016). Public participation, land use and climate change governance in Thailand. *Land Use Policy, 52*, 511–517. https://doi.org/10.1016/j.landusepol.2014.12.014

Kaufmann, D., Kraay, A., & Mastruzzi, M. (2011). The worldwide governance indicators: Methodology and analytical issues. *Hague Journal on the Rule of Law, 3*(02), 220–246. https://doi.org/10.1017/s1876404511200046

Keskitalo, E. C. H., & Kulyasova, A. A. (2016). The role of governance in community adaptation to climate change. *Polar Research, 28*(1), 60–70. https://doi.org/10.1111/j.1751-8369.2009.00097.x

Mees, H., & Driessen, P. (2018). A framework for assessing the accountability of local governance arrangements for adaptation to climate change. *Journal of Environmental Planning and Management, 62*(4), 671–691. https://doi.org/10.1080/09640568.2018.1428184

Ministry of Environment, Forest and Climate Change. (2009a). *Bangladesh climate change strategy and action plan 2009.* MoEFCC, Government of the People's Republic of Bangladesh.

Ministry of Environment, Forest and Climate Change. (2009b). *National adaptation programme of action.* https://unfccc.int/resource/docs/napa/ban02.pdf

Rahman, M. M., & Huang, D. (2019). Climate governance initiatives: Snapshots from Bangladesh. *Journal of Geoscience and Environment Protection, 7*(9), 131–147. https://doi.org/10.4236/gep.2019.79010

Sarzynski, A. (2015). Public participation, civic capacity, and climate change adaptation in cities. *Urban Climate, 14*, 52–67. https://doi.org/10.1016/j.uclim.2015.08.002

Sippel, M., & Jenssen, T. (2009). What about local climate governance? A review of promise and problems. *Social Science Research Network.* https://mpra.ub.uni-muenchen.de/20987/

Sprain, L. (2016). Paradoxes of public participation in climate change governance. *Good Society, 25*(1), 62–80.

United Nations Development Programme. (2016). *Human development report 2016: Human development for everyone.* http://hdr.undp.org/sites/default/files/2016_human_development_report.pdf

Vivekanada, J. (2013). *Governance—The key to climate resilience in Bangladesh?* http://www.international-alert.org/blog/governance-key-climate-resilience-bangladesh

Chapter 7
Conclusion and Recommendations

7.1 Summary

This research has examined the climate change governance practices of GOs and NGOs in Bangladesh. This research empirically assessed community perceptions of climate change governance practices in Hatiya Upazila, as the selected study area. This area has experienced a number of climate change-induced disasters in recent years. This research also examined the performances of GO and NGOs in climate governance practices. Questions were developed and examined to evaluate the nature and extent of existing climate change governance practices, the roles of different stakeholders in climate governance, the existing challenges in climate governance, and recommendations to overcome these challenges and make current climate change governance more effective.

Following an MMR approach, data were collected through head-of-household surveys, KI interviews, FGDs, a transect walk and secondary data sources. The application of MMR facilitated the systematic collection of quantitative and qualitative data through data source triangulation, and provided a more complete understanding of the research issues raised by this study.

Chapter 3 outlined, along with the research methods, the geographical position of the study area in the coastal part of Bangladesh, providing a detailed description of its sociopolitical setting. This area differs from other parts of the country in many respects, such as local vulnerability and the local people's socioeconomic conditions. These were assessed through both fieldwork and other secondary sources. Hatiya Upazila, the present study area, has been affected by frequent climate change-related disasters and has a high level of climate vulnerability. It was therefore an ideal location to undertake the present research on climate change governance.

Chapter 4 described the roles of GOs in climate change governance practices. GOs play the lead role in planning and implementing various climate change-related plans and programs. GOs have developed a number of plans and strategies addressing climate change Bangladesh. The government has integrated various procedures to

Md. F. Hossain and S. Momtaz, *Evaluating Climate Change Governance in Bangladesh*,
https://doi.org/10.1007/978-3-031-98486-0_7

meet basic governance criteria, such as accountability, transparency, participation, government effectiveness and resources, but these procedures do not meet the expectations of climate-vulnerable people owing to the misuse of resources and corrupt practices at the government level. Policy development and implementation processes are largely based on a top-down approach. The government is nevertheless working in different ways to mainstream climate change issues and make an effective change in this sector.

Chapter 5 highlighted the climate change governance practices of NGOs in Bangladesh. NGOs are an inseparable part of overall development in Bangladesh. They work on climate change issues from a national to the local level, and give important impetus to the planning and implementation of climate change-related plans and programs. They also follow governance protocols while working on climate change-related issues. NGOs, through their activities on climate change, are having a positive impact on vulnerable people's lives and livelihoods. Although the surveyed NGOs face various challenges in climate change-related activities, their procedures and protocols play an important role in improving climate governance initiatives. NGOs work at different levels on many climate change issues to protect people's lives and livelihoods from the adverse effects of climate change.

Chapter 6 presented community perceptions of climate change governance. Five governance criteria—accountability, transparency, participation, government effectiveness and resources—were examined to evaluate community perceptions of climate change governance. This study demonstrated the generally low level of local community participation in climate governance. Overall, community perceptions and views were frequently not considered when planning and implementing various climate projects at the local level. Community participants claimed that they did not receive the required support in climate change-related issues from local administrations, because of the limited capacity to solve climate change-related issues, even though they were affected significantly by climatic events. Several initiatives relating to climate change issues were observed nationally, but at the local level, climate governance does not function well or is non-existent.

Good governance is a prerequisite for a country's rapid economic growth and development. Climate change governance is vital as a subset of governance. If the entire governance mechanism is ineffective, it is difficult to achieve good climate governance. In contrast, if the entire governance mechanism is robust, it will be easier to achieve good climate change governance. In the context of Bangladesh, governance has both positive and negative aspects. Improving existing governance systems by building on positive elements while simultaneously addressing negative aspects is a significant challenge. In that context, climate governance has an important role to play in improving overall governance, through measures such as ensuring the accountability of relevant stakeholders, enhancing the transparency and effectiveness of investments, and encouraging full public participation.

7.2 Recommendations

The findings of this research raise several contextual limitations regarding climate change governance practices, from the perspective of GOs and NGOs in Bangladesh. These findings indicate challenges in the current plans, programs, funding strategies and implementation processes relating to climate governance. Based on this investigation, recommendations are given here, to create a more effective climate change governance system by improving the existing planning, implementation, training and procurement processes. This should lead to an environment more conducive to increasing resource mobilisation and the proper implementation of various climate projects. If significant funding is made available for climate projects, but is then not spent effectively with good governance, poor or no outcomes may result. This may result in increasingly difficulty obtaining future funding climate-vulnerable countries such as Bangladesh. Based on its findings, this study makes the following recommendations:

1. The majority of government officials, working in the MoEFCC and line ministries, do not have an adequate knowledge of climate change that will equip them to deal with complex projects at the government level. Government must therefore enhance the capacity of relevant agencies and officials to improve and review adaptation and mitigation plans, develop good projects to secure required funds and ensure other projects already recommended in climate change documents, such as NAPA, BCCSAP, are implemented. Capacity development of government officials must be undertaken at national, regional and local levels.
2. Bangladesh must develop a separate independent national body, such as an independent climate change commission, to oversee climate change-related issues and programs. All planning, funding and programs relating to climate change should be brought under one umbrella, ensuring better coordination, integration and accountability of all climate programs and projects. This body should develop a plan that better prioritises and integrates all climate change-related works. They should also develop a research platform to enhance research activities on climate change issues. Such a body would avoid the duplication of climate projects and provide for effective climate governance. For example, the Climate Change Commission in the Philippines is the lead policy-making body at the government level, run by a vice-president who is directly responsible to the president. This type of oversight body is required in Bangladesh to manage, monitor, engage and expedite activities relating to climate change.
3. Stakeholder involvement, accountability and transparency must be ensured in the climate governance system. Result-based monitoring is essential to obtain better results from climate projects. Stakeholder participation will also ensure the proper implementation of climate projects. The corruption and mismanagement currently inherent in climate change projects should be curtailed through stakeholder involvement, which will lead to better outcomes. Stakeholder involvement is the best way of ensuring that the pledges are met, as they will have a genuine stake in the program outcomes ongoing maintenance.

4. Most plans on climate change issues are developed centrally, so these should also consider bottom-up approaches involving think tanks and research organisations working collaboratively on climate change issues. The existing policies relating to climate change should be revised to ensure a bottom-up approach through stakeholder participation, to ensure increased effectiveness.
5. Government departments such as BCCT, IMED and CAG should be modernised so they have the required professional and technical capacity to oversee, monitor or audit (as required) the finances and outcomes of climate change programs and projects. Two types of audit are important: during a project and post-project. Watchdog bodies, such as TIB, have a professional capacity for monitoring and auditing, and should be engaged by the government as independent bodies to oversee appropriate projects.
6. Government ministries and departments are working on climate change issues because abundant funding has, at times, been made available. These programs often overlap. Conflict with programs run by GOs, and even sometimes with NGOs can arise, leading to poor outcomes and minimal benefits. Such practices should be stopped and proper monitoring established. Recommendation 2 above goes some way to address this problem, but a whole-of-government approach is needed to overcome such inefficiencies.
7. Various stages of climate change programs and projects, such as planning (Recommendation 1), capacity development (Recommendation 5), procurement, stakeholder involvement (Recommendation 9) and implementation, should be improved and made transparent. Wider access to climate funds should be available, along with improved governance in the use of such funds. If today's climate change funds are not used properly, with good governance, then it will be very difficult to obtain funds in the future, even though the importance of and urgency for funding is bound to increase.
8. The government needs to scale up knowledge, capacity, research and involvement on climate change issues at the local level. Local-level government institutions should increase their capacity to deal with climate change, and develop proper plans, programs and implementation processes concerning climate change activities at the local level. Local administration offices or CBOs should be involved in oversight systems, taking responsibility for climate project funds. The ever-increasing funds dedicated to climate projects make no sense unless proper fund use is ensured. At the *upazila* and union levels, coordination gaps need to be addressed, and transparency, accountability and participation ensured to maximise the benefits arising from such projects; in other words, 'improved governance at the local level'.
9. Community stakeholders should be involved in climate projects so they can receive greater benefits from climate change-related activities. Community participation is needed to ensure the quality of work, and to make projects more effective. Community stakeholders also should support development activities in their area to reduce the adverse effects of climate change. If climate change projects are implemented properly, with adequate consideration of local settings

and local knowledge, then climate change governance practices are likely to bring improved community resilience to face climate change.

10. To make climate change-related initiatives more successful, focus should be placed on introducing effective 'local intermediaries', who would be responsible for using climate funds more efficiently and disbursing them more effectively.
11. NGOs have an important role to play in the climate change governance system, because they frequently work in parallel with government departments in addressing climate change. NGOs should therefore play a major role in the establishment, implementation and monitoring of any future climate change governance protocols and procedures.
12. NGOs face difficulties at the local level while they work on climate change projects. They need support from local administrations in their activities relating to climate change. Local administrations should have the capacity to differentiate between climate funds and development funds, and be given responsibility to work with NGOs when there are difficulties with local programs regarding these designations.
13. Climate change-related activities are often affected by political interference. It is therefore vital to remove any pressure from political representatives. Climate project approval through political interference should be stopped to ensure project quality and integrity. Nevertheless, politicians have important roles to play in climate governance. Climate governance policies and practices should be formulated with the support of political representatives to reduce ruling party dominance. Ministerial oversight and effective political input at the planning stage of climate change policies and programs can enhance the proper use of existing resources to combat climate change and better address climate change-induced disasters.
14. Effective measures should be taken to improve climate change governance at all levels. GOs cannot implement all tasks alone, but they should play a proactive role in undertaking effective collaboration and partnership with donors, international organisations, NGOs, CBOs and the community.

7.3 Study Contributions

Initially, this research developed a conceptual framework after reviewing the literature on climate change governance and the main criteria for evaluating climate governance practices. This research describes how different stakeholders are dealing with climate change governance practices in their respective domains, with focus on GOs, NGOs and community-levels. The findings relating to these are illustrated in Chaps. 4–6.

Considering that very few studies on climate change governance currently exist, this research is a significant addition to the current literature on these issues relating to Bangladesh. The findings of this study are relevant to the plans and strategies for addressing the effects of climate change in Bangladesh. The research results provide significant contributions towards climate change governance-related research and assessments including the collection, analysis and reporting of climate change data to improve our understanding of the complex issues involved. This study makes recommendations for improving climate change governance practices, including improvement in relevant stakeholder performance, maximising resource use and ensuring the involvement of critical activities at the community level. While this study critiques current government initiatives on climate change governance practices, it ultimately recommends how to improve current governance protocols and procedures to ensure their effectiveness. This research emphasises the vital role of local communities in climate change programs and governance practices: the people who are the actual victims of climate change-induced disasters. Ensuring the involvement of various stakeholders in project implementation processes can make these climate change-related projects more successful.

The MMR approach, in association with pragmatism, has provided methodological, theoretical and philosophical aspects to evaluate the spectrum of climate governance issues from the national to the local level. This approach is an effective research technique to combine qualitative and quantitative data. This approach helped to summarise the research findings through several themes and sub-themes, providing understanding and explaining different issues regarding climate change governance practices, the role of GOs and NGOs in climate governance, challenges in climate governance, community perceptions of climate governance, the effectiveness of climate governance and stakeholders' most anticipated features of effective climate governance. This approach enabled conceptualisation of climate governance issues and demonstrated the study's empirical findings systematically.

To the best of the researcher's knowledge, no other study exists that has evaluated climate change governance in Bangladesh, particularly one using empirical findings. From that perspective, this is a unique research project on climate change governance in Bangladesh. The proposed recommendations provide a framework for Bangladesh's government to not only improve climate change governance practices, but also promote Bangladesh's SDGs.

7.4 Study Limitations

This research followed a rigorous methodology to obtain detailed insights and to ensure reliability and validity, but a few limitations were observed. The limitations include time constraints, bureaucratic procedures for gaining approval from relevant authorities while collecting data, accessibility to study sites on a remote island, and the scheduling (and rescheduling) of interview times and dates for respondents, because of their involvement in other activities. Before beginning fieldwork, ethics and safety protocols were followed to maintain research validity and reliability.

This research was conducted based on a MMR approach, comprising qualitative and quantitative data, followed by other data collection procedures. Phases of data processing included data entry, transcription, translation, and the coding and analysis of both types of data. Compiling the empirical findings was very time-consuming. Managing a large number of head-of-householder surveys, KIIs and focus group discussions to obtain the expected fieldwork outputs was a rigorous task. Although time-consuming, the researcher managed such difficulties with the utmost care.

This study only considered the perceptions of selected communities, and these may not represent those existing elsewhere in the country. This study presented its findings on the local scale of one *upazila*. The sample size was determined by following rigorous statistical methods. Owing to time limitations, it was not possible to incorporate a larger number of respondents to obtain better outputs. The study does, however, clearly demonstrate the field-level perceptions of climate governance practices according to the surveyed communities.

It took approximately four months to conduct the fieldwork for this research in Bangladesh. Before starting the surveys at the community level, permission was sought from the local administration. Permission was provided by various GOs and NGOs to interview the nominated representatives of those organisations. Fieldwork was scheduled around the general election time when the government was in transition. This may have affected some responses from government departments, as policy positions were not yet fixed for the incoming government. The comprehensive research literature was therefore used and precautions were taken throughout the research to minimise any shortcomings.

7.5 Future Research Directions

This research is focused on evaluating the existing climate change governance practices of GOs and NGOs in Bangladesh. Community perceptions were also considered to examine climate governance practices at the local level and to validate the research findings. This research links previous research on environmental governance, as climate governance is highly associated with this.

Future complementary research directions could focus on six areas.

First, climate change governance is evolving in Bangladesh. Similar research should be undertaken to assess the socioeconomic and environmental effects of climate change governance and its impact on the country's overall development. Such research could empower stakeholders to promote the input and coordination of their activities on climate change issues.

Second, future research should explore legal issues associated with climate change governance. By revising and updating the present legal instruments relating to climate change, vulnerable communities could be given improved powers and involvement, while the capacities of authorities involved in climate activities (especially at the local level) will be increased. This is highly interrelated with the context of Sustainable Development Goal 13 ('Take urgent action to combat climate change and its impacts'), with its emphasis on undertaking climate change-related activities more efficiently.

Third, this research followed five criteria (e.g. accountability, transparency, participation, government effectiveness and resources) to evaluate climate governance practices of GOs, NGOs and communities. Future research could be conducted encompassing all other criteria such as the rule of law, political stability, equity, institutional capacity and inclusiveness, leading to a better understanding of the effectiveness of other criteria on government practices.

Fourth, climate change governance practices are linked strongly to environmental governance in Bangladesh. Future research could focus on how climate governance and environmental governance can be better integrated and intertwined for improved outcomes.

Fifth, this research was conducted in a particular area to gain in-depth insights from local communities. There are many climate change-vulnerable areas in Bangladesh where climate change issues are an integral part of the lives and incomes of local people. Various parts of the country are facing climate change-related problems in different ways in their differing local contexts. Future research could explore issues around climate governance in other climate-vulnerable parts of the country more effectively. Based on such findings, improved communication strategies could be developed for those particular communities, which could enhance their resilience to tackle climate change effects specific to their needs and priorities.

Sixth, research could be conducted to assess how much has actually been achieved by the various climate projects of different stakeholders. Research could focus on loopholes in the governance processes and protocols in particular climate projects and how these projects addressed on-ground outcomes relating to climate change governance issues from their beginning to end.

Appendix A
Head-of-Household Survey Semi-structured Questionnaire

Date of interview			
Area			

Evaluating Climate Change Governance in Bangladesh—Perspective of Governmental and Non-governmental Organizations

[Data collected for this research was used for research purposes only. Anonymity and confidentiality were maintained strictly. No information was disclosed to anyone except the researcher and supervisor]

For this research, an evaluation was conducted on the climate change governance practices of governmental and non-governmental organizations in Bangladesh. A series of semi-structured questions were asked to conduct this head-of-household survey on the following issues in relation to the main topic.

Please put a tick (√) mark on the box that represents your answer.

Md. F. Hossain and S. Momtaz, *Evaluating Climate Change Governance in Bangladesh*,
https://doi.org/10.1007/978-3-031-98486-0

Part A: Basic Information

A01: Head-of-Household survey no: ……../……….
A02: Gender: □ Female □ Male
A03: Age range: □ 18–19 □ 20–29 □ 30–39 □ 40–49 □ 50–59 □ 60+
A04: Marital status: □ Single □ Married □ Divorced □ Widowed □ Separated
A05: Education level:

1 = Cannot read and write	
2 = Primary	
3 = Secondary	
4 = Technical/vocational	
5 = College	
6 = University	

A06: Religion 1 = Muslim, 2 = Hinduism, 3 = Buddhist, 4 = Christian, 5 = Others
A07: Occupation 1 = Farming, 2 = Fishing, 3 = Trade, 4 = Job, 5 = Retirement, 6 = Others
A08: Household size:
1 = 0–2
2 = 3–4
3 = 5–6
4 = More than 6
A09: Residential house:
1 = Pucca (brick, cement, iron)
2 = Semi-pucca (brick wall, tin roof)
3 = Kutcha/jupri (mud, bamboo, leaves)

Part B: Climate Change

B01: Have you ever heard of climate change?  Yes  No

B02: What is climate change? (Tick as many as you wish)

It is weather condition	o
More extreme weather events, such as cyclone, storm surge, flood etc.	o
More rainfall	o
Less rainfall	o
Extreme cold in winter	o
More days with high temperature	o
Unpredictable weather condition	o
Long-term shifts of weather patterns	o
Others	o

B03: What kind of climatic events are observed in your area? (Please indicate your level of observation)

Events	Very low	Low	Moderate	High	Very high
Flood					
Cyclone					
Drought					
Coastal erosion					
Storm surge					
Salinity intrusion					
Others					

B04: Do you think climate change has been affecting you as an individual?

1. Yes 2. No 3. No response (If Yes, then go to the question no B05)

B05: What are the effects of climate change on you and your livelihoods?

Increased flooding/cyclone/storm surge/salinity intrusion/drought	o
Increase rainfall	o
Unpredictable rainfall pattern	o
Stronger winds	o
Fresh water shortage	o
Warmer summers	o
Cooler winter	o
Income decreases	o
Loss of crops	o
Loss of property	o
Disruption of children's education	o
Loss of employment	o
Displacement of people	o
Poverty increases	o
Increased incidences of diseases	o

Part C: Climate Change Governance Practices

C01: Do you know about details of climate change projects in your area?

1. Very poor 2. Poor 3. Medium 4. Good 5. Very good

o o o o o

C02: Are government organizations working on various climate change related issues?

1. Strongly disagree 2. Disagree 3. Medium 4. Agree 5. Strongly agree

o o o o o

C03: Are non-government organizations working on various climate change related issues?

1. Strongly disagree 2. Disagree 3. Medium 4. Agree 5. Strongly agree

o o o o o

C04: How often do you visit climate change related projects in your area?

1. Never 2. Rarely 3. Monthly 4. Fortnightly 5. Weekly

o o o o o

C05: How often do you visit government office working on climate projects in your area?

1. Never 2. Rarely 3. Monthly 4. Fortnightly 5. Weekly

o o o o o

C06: How often do you visit non-government organizations (NGOs) working on climate projects in your area?

1. Never 2. Rarely 3. Monthly 4. Fortnightly 5. Weekly

o o o o o

C07: How often the government officials visit climate projects in your area?

1. Never 2. Rarely 3. Monthly 4. Fortnightly 5. Weekly

o o o o o

C08: How often the non-government officials visit climate projects in your area?

1. Never 2. Rarely 3. Monthly 4. Fortnightly 5. Weekly

o o o o o

C09: How do you rate the relationships with various stakeholders relating to climate governance?

1. Very poor 2. Poor 3. Medium 4. Good 5. Very good

o o o o o

C10: Are you currently engaged in any form of climate governance related committee?

1. Yes 2. No 3. No response

C11: Do you attend any meeting at any level in existing climate change related works?

1. Yes 2. No 3. No response

C12: If you attended any meeting, do you take part in discussion and give your feedback?

1. Yes 2. No. 3. No response

C13: How would you rank your participation in climate governance practices in terms of your engagement and contribution?

1. Very poor 2. Poor 3. Medium 4. Good 5. Very good

o o o o o

C14: Do you think the previous activities relating to climate change have maintained participation of all relevant stakeholders?

1. Strongly disagree 2. Disagree 3. Medium 4. Agree 5. Strongly agree

o o o o o

C15: Transparency is maintained in climate change related projects?

1. Strongly disagree 2. Disagree 3. Medium 4. Agree 5. Strongly agree

o o o o o

C16: Information relating to climate projects are shared with the people in your area?

1. Strongly disagree 2. Disagree 3. Medium 4. Agree 5. Strongly agree

o o o o o

C17: Were the climate change related projects completed on time?

1. Strongly disagree 2. Disagree 3. Undecided 4. Agree 5. Strongly agree

o o o o o

C18: Did the climate change related projects address the problems of local people properly?

1. Strongly disagree 2. Disagree 3. Undecided 4. Agree 5. Strongly agree

o o o o o

C19: Any priority setting relating to climate projects was maintained in your area?

1. Strongly disagree 2. Disagree 3. Undecided 4. Agree 5. Strongly agree

o o o o o

C20: How would you rank the successes of climate projects in your area?

1. Very good 2. Good 3. Undecided 4. Poor 5. Very poor

o o o o o

C21: Who is the most important player in climate change governance systems?

1. GOs 2. Political leaders 3. NGOs 4. Community leaders 5. Community people 6. Others

o o o o o o

C22: Do you think the climate projects have been implemented properly?

1. Strongly disagree 2. Disagree 3. Undecided 4. Agree 5. Strongly agree

o o o o o

C23: How would you rate the quality of your relationships with climate governance actors and institutions?

Officials/actors	Relationships and community output				
	1. Very weak	2. Weak	3. Medium	4. Strong	5. Very strong
1. Government officials	o	o	o	o	o
2. NGO officials	o	o	o	o	o
3. Community based organizations	o	o	o	o	o
4. Community people	o	o	o	o	o

D23: Did the local government departments take collective strategies/actions when climate change related problems happened?

Yes = 1; No = 2; No answer (If answer is 'Yes', then go to question no D24)

D24: What are the collective strategies or actions for climate change problem?

....

D25: Did the community people take collective strategies/actions when climate change related problems observed?

Yes = 1; No = 2; No answer (If answer is 'Yes', then go to question no D26)

D24: What are the collective strategies or actions for climate change problem?

....

Part D: Perception About Government Effectiveness

Please put a tick (√) mark on the option that best represents your perception about climate change governance.

S No.	Specific opinion item	Strongly agree	Agree	Neither agree or disagree	Disagree	Strongly disagree
1	Presence of institutional setting for climate governance	o	o	o	o	o
2	Policy and plans are in my favour	o	o	o	o	o
3	Participation of various stakeholders are ensured	o	o	o	o	o
4	Personal rights and responsibilities in climate governance practices are not well defined	o	o	o	o	o
5	Adequate funds for climate projects	o	o	o	o	o
6	Accessibility in climate change projects	o	o	o	o	o
7	Stakeholders' capacity building	o	o	o	o	o
8	Institutional capacity building	o	o	o	o	o
9	Timely completion of climate change projects	o	o	o	o	o
10	Regular monitoring of climate change projects	o	o	o	o	o
11	Climate change governance system is effective	o	o	o	o	o
12	Climate change projects ensures the involvement of locals	o	o	o	o	o
13	Rules and policies are well informed	o	o	o	o	o
14	Climate projects involve locals in decision-making processes	o	o	o	o	o

(continued)

(continued)

S No.	Specific opinion item	Strongly agree	Agree	Neither agree or disagree	Disagree	Strongly disagree
15	Existing climate projects are supportive for local livelihoods	o	o	o	o	o
16	Climate projects are enhancing public awareness	o	o	o	o	o
17	Climate projects are making significant positive impacts on local livelihoods	o	o	o	o	o
18	Female members' participation are encouraged in various projects	o	o	o	o	o
19	Local interests are prioritized and emphasized	o	o	o	o	o
20	Employment opportunities for locals have not been created	o	o	o	o	o
21	Vulnerability and uncertainty of livelihood have not decreased	o	o	o	o	o
22	Confrontations are observed among GOs and NGOs in climate change related projects	o	o	o	o	o
23	Climate change governance has changed over time	o	o	o	o	o
24	Effective climate change governance is expected	o	o	o	o	o

Appendix B
Questionnaire for the Key Informants

Key informants to be interviewed:

I. Community people
II. Government officials (national and local level)
III. Representatives of NGOs (national and local level)
IV. Civil society representatives-academics, researchers and journalists.

[Data collected for this research was used for research purposes. Anonymity and confidentiality were maintained strictly. No information was be disclosed to anyone except the researcher and supervisor]

For this research, an evaluation was conducted on the climate change governance practices of governmental and non-governmental organizations in Bangladesh. A series of semi-structured questions were asked to conduct this key informant interview on the following issues in relation to the main topic.

Introduction:

Introduce myself, the research, establish ethical consent

Can you please tell us about your role within your organization?

1. What is the role of your organization in climate change issue?
2. Have the climate change issues been incorporated into plans and programs of your organization?
3. What are the successes from the previous and current climate change related projects?
4. What are the present and future challenges your organization facing in implementing the projects?
5. How do you maintain accountability of your organization on climate change related issue?
6. How do you ensure participation of the relevant stakeholders in climate change related projects?
7. How transparency is maintained in various climate change related activities?

Md. F. Hossain and S. Momtaz, *Evaluating Climate Change Governance in Bangladesh*, https://doi.org/10.1007/978-3-031-98486-0

8. Do you get adequate support from other agencies to conduct the climate change related projects?
9. What are the sources of funding for various climate change related projects?
10. How do you allocate the funds in your organization in various climate change related projects?
11. What is the institutional framework for climate change governance in your organization?
12. What are the monitoring system for climate change related projects?
13. According to you, has climate change governance changed over the time? What are the changes?
14. What are the successes of climate change governance system of your organization?
15. What are the challenges for climate change governance of your organization?
16. What suggestions would you like to make for an effective climate governance?
17. Do you have any other comments about climate change governance that has not been covered today that you would like to share?

Thank you for your cooperation

Appendix C
Questions for Focus Group Discussions

[Data collected for this research will only be used for research purposes. Anonymity and confidentiality will be strictly maintained. No information will be disclosed to anyone except the researcher and supervisor]

For this research, I will evaluate the climate change governance practices of governmental and non-governmental organizations in Bangladesh. I will ask a series of semi-structured questions to conduct this focus group discussion on the following issues in relation to the main topic.

1. What kind of climatic events (flood, cyclone, coastal erosion, storm surge, salinity intrusion, drought etc.) are observed in your area?
2. Have you noticed any changes of climate in last 20 years?
3. What are the consequences/impacts of climate change on your livelihoods?
4. Which government organizations (GOs) and non-governmental organizations (NGOs) are working on climate change in your area?
5. How are GOs working for climate change in your area? What are the perceptions of community people about GOs' activities?
6. How are NGOs working for climate change in your area? What are the perceptions of community people about NGOs' activities?
7. What are the roles of community people in climate change activities? Are their participations important? How?
8. What are the changes of climate governance over time in this area?
9. What are the successes of GOs in climate change related activities?
10. What are the successes of NGOs in climate change related activities?
11. What are the challenges for GOs in climate change issue in your area?
12. What are the challenges for NGOs in climate change issue in your area?
13. What suggestions would you like to make for an effective climate change governance?

Md. F. Hossain and S. Momtaz, *Evaluating Climate Change Governance in Bangladesh*,
https://doi.org/10.1007/978-3-031-98486-0

14. Do you have any other comments about climate change governance that hasn't been covered today that you would like to share?

Thank you for your cooperation

Appendix D
Frequently Used Words in KIIs and FGDs (Word Clouds)

Md. F. Hossain and S. Momtaz, *Evaluating Climate Change Governance in Bangladesh*,
https://doi.org/10.1007/978-3-031-98486-0

The manufacturer's authorised representative in the EU is Springer
Nature Customer Service Centre GmbH, Europaplatz 3, 69115 Heidelberg,
Germany. If you have any concerns regarding our products, please
contact ProductSafety@springernature.com

Printed and bound by CPI Group (UK) Ltd, Croydon, CR0 4YY
30/09/2026
02225588-0002